T0136717

Being Young, Male and Muslim in Luton

SPOTLIGHTS

Series Editor: Timothy Mathews, Emeritus Professor of French and Comparative Criticism, UCL

Spotlights is a short monograph series for authors wishing to make new or defining elements of their work accessible to a wide audience. The series provides a responsive forum for researchers to share key developments in their discipline and reach across disciplinary boundaries. The series also aims to support a diverse range of approaches to undertaking research and writing it.

Being Young, Male and Muslim in Luton

Ashraf Hoque

First published in 2019 by
UCL Press
University College London
Gower Street
London WC1E 6BT

Available to download free: www.ucl.ac.uk/ucl-press

A CIP catalogue record for this book is available from The British Library.

ISBN: 978–1-78735–136-3 (Hbk.)
ISBN: 978–1-78735–135-6 (Pbk.)
ISBN: 978–1-78735–134-9 (PDF)
ISBN: 978–1-78735–137-0 (epub)
ISBN: 978–1-78735–138-7 (mobi)
ISBN: 978–1-78735–139-4 (html)
DOI: https://doi.org/10.14324/111.9781787351349

For my mother,
My mother,
My mother,
And then my father.

Contents

Acknowledgements

This book has been in process for some time, and would not have been concluded without the generous support of so many. I will be forever grateful to all of them. In particular, I would like to thank my PhD supervisor, Parvathi Raman, for her ceaseless encouragement, wisdom, and patience. I would also like to thank my teachers at SOAS Anthropology and History. Specifically, William Gervais Clarence-Smith, John Parker, Magnus Marsden, Richard Fardon, Jakob Klein, Dolores Martinez, Kostas Retiskas, Caroline Osella, George Kunnath, Christopher Davis, Kevin Latham, David Mosse, Harry West, Gabriele vom Bruck, Trevor Marchand, Edward Simpson, Paul Francois-Tremlett and Cosimo Zene from the Study of Religions department. At UCL, I owe a huge debt of gratitude to Lucia Michelutti, Michael Stewart, Martin Holbraad, Allen Abramson, Charles Stewart, Lewis Daly, Alison Macdonald, Guiherme Heurich, and Dalia Iskander for their selfless guidance and blind faith throughout the final stages. Lastly, to Arild Ruud for gently pushing me off the plank.

This work was conceived, somewhat naively, with the help of my two dear friends, Hasan Al-Khoee and Igor Cherstich. I would like to thank them, along with the Brotherhood of Truth and Justice [and Gratitude] (you know who you are). A huge thank you to the people of Bury Park for their boundless hospitality, insight, and humour. Most notably, my friends at the MSP for hiring, hosting, and educating me. A work like this is so difficult to do justice to, especially given the duty to represent friends and interlocutors with necessary sensitivity and accuracy. Even so, errors never announce themselves. It thus goes without saying, as ever, that all such mistakes are my own alone.

Finally, I would like to extend my eternal gratitude to my family for their unconditional support throughout my life. My wife, Rosie, was a great source of relentless kindness and sustenance, for which I can never really reimburse her. Most profoundly, however, I would like to acknowledge my late parents. Principally, the tireless efforts of my mother in raising me against all odds, and demonstrating that anything is possible.

Introduction

This book will seek to provide an anthropological account of the lives of young British-born Muslim men of South Asian origin in the English town of Luton. Luton is a satellite town in close proximity to London, situated around thirty miles to the north of the city. It is an important town. Many of its inhabitants work in London, which is a convenient forty minutes' train journey away. Luton also boasts an international airport that plays a significant part in providing air-travel needs for most of southern and central England. The town is further served by the M1 motorway, which connects London with Leeds in the previously industrial north of the country. In a previous life, Luton was an industrial town in its own right: famed for its manufacture of hats and, later, commercial and domestic motor vehicles. Since the turn of the century, however, these industries have all but vanished with the manufacturing industry being steadily replaced by the services sector – a phenomenon consistent with national trends. Historically speaking then, Luton is not a remote town. Its sophisticated transport links, coupled with once thriving industry, has encouraged people from up and down the country to pass through, work and settle in its environs. Among these settlers is a substantial Muslim community – around 19% of the overall population of the town which is approximately 200,000 people. These Muslims predominantly reside in a concentrated neighbourhood situated in and around the Bury Park area of the town. Like many of the town's inhabitants, they were attracted to Luton during its industrial zenith, taking up employment in the Vauxhall Motor Company, its brickyards and engineering factories. Over the years, Muslim residents have burgeoned into a sizeable settled community where new generations, who were born and raised in the town, now call it their 'home'. My fieldwork was conducted in this post-industrial landscape among this new generation of Britons. The bulk of it was conducted between 2008 and 2010, with further observational visits undertaken in the years since.

The chief concern of this book is to demonstrate, through observations of their everyday lives, the various ways British-born Muslim men in Luton develop understandings of themselves that transcend the monolithic contemporary image of British Muslim communities. I suggest that young Muslims in Luton are developing hybridised forms of Islam in an attempt to find alternatives to the culturally exclusive state narratives of nationality and citizenship. I argue that the syncretic experiences within the habitus of the South Asian Muslim 'home', coupled with influential interactions with 'mainstream British society', has led to the rejection of both points of reference. Rather, young British Muslims are looking to Islam – its 'glorious past' and the international Muslim community (*ummah*) – as a means of reconciling both the alienating cultural practices of the home and the perceived Islamophobia and defilement of society beyond. By resorting to an abstract re-articulation of Islam, young Muslims are re-adapting their communal affiliations to suit the social and cultural terrain of Britain, while simultaneously creating a counter-discourse to accepted notions of 'Britishness' and national belonging. Although a substantial and credible corpus of sociological and anthropological literature has emerged in recent decades seeking to situate and problematise British Muslim communities, this study will further illuminate the inherent complexities and resultant 'messiness' of this task. Furthermore, I aim to provide an analysis of Muslims in Britain that attempts to depart from certain conceptions that view diaspora Muslims as a separate cultural bloc, and move towards an understanding that identities are historically fluid, constantly in flux and continuously shifting social categories and markers.

***

I have chosen to focus this account on young Muslim men for two reasons. Firstly, I had relatively more access to men than I had to women. Although I did interact with and interview women (and some of their insights are implicit within the analysis), I was mindful of cultural sensitivities pertaining to mixed-gendered interactions. Although women were generally keen to partake in the research, participant observation was limited to schools, colleges, and other public spaces, mostly during the daytime. On the other hand, I was able to access male informants in both private and public spaces, to share residential trips with them, and to stay in their homes. Secondly, while much anthropological attention has been bestowed on Muslim women in diaspora, particularly with regards to the *hijab* debate[1] there is far less attention on Muslim men going about their everyday lives in the West.[2] I find this group particularly interesting as, like the 'hyper-masculine' inner city black males before them,[3] they have increasingly become categorised as the archetypical 'folk devil' by bourgeois society. Claire Alexander (2000) argues that the public perception of Muslim men reflects 'a growing concern with the "problem" of Asian youth – and more specifically, with the problem of *Muslim young men*.[4] If they share the same well-established tropes of racial alienation and social breakdown that created, and continue to create, moral panics of Rastafarian drug dealers, black rioters, muggers and Yardies [...], what they also reflect is a new cultural formation' (Alexander 2000:4 original emphasis). A year after her book was published, young Muslim men of Bangladeshi and Pakistani heritage made national news after race riots in the former industrial northern mill towns of Oldham, Burnley and Bradford, in which 200 police officers were injured. According to Arun Kundnani, these riots represented an out-spilling of resentment after decades of state-endorsed segregation along ethnic and racial lines:

> By the 1990s, a new generation of young Asians, born and bred in Britain, was coming of age in the northern towns, unwilling to accept the second-class status foisted on their elders. When racists came to their streets looking for a fight, they would meet violence with violence. And with the continuing failure of the police to tackle racist gangs, violent confrontations between groups of whites and Asians became more common (Kundnani 2001:108).

Negative perceptions of Muslim men were, of course, compounded by the events of 7/7, when four British Muslims, three of whom were of Pakistani origin, raised in Yorkshire, were posthumously found to be responsible for the attacks. The image of the male Muslim 'folk devil' was further crystallised by various all-male Muslim grooming gangs operating during the night-time economy in towns and cities across the country. Most recently, London and Manchester have witnessed a sequence of terror attacks orchestrated and executed by young British Muslim men. In addition to these, the British press has, since the declaration of the 'war on terror', been consistently reporting on Muslim men travelling to Syria to join ISIS, plots to detonate homemade bombs in public places, and ideological 'hate preachers' running British mosques.[5] This book, therefore, is an attempt to explore the lives of young Muslim men with these associated connotations in mind. Furthermore, I found that not only were young men acutely aware of these wider perceptions, but many engaged in sardonic and ironic performances in affirmation of such stereotypes. 'If you call someone an extremist for long enough, he might just become one', one of my key informants reminded me. Thankfully, I did not come across many would-be or actual terrorists during my time in the field. I did, however, meet with plenty of young men deeply annoyed and emotionally saturated by derogatory impressions of them in the public sphere. Almost twenty years on from Claire Alexander's *The Asian Gang*, young Muslim men remain, it seems, a social 'problem'.

One area where a sense of masculinity was particularly pronounced was in the realm of making money. My informants were committed to securing livelihoods that yielded maximum financial rewards. They were often encouraged to get a job and secure employment as soon as they possibly could by parents and grandparents. Once in a job, wages and profits were habitually funnelled into the family purse. The more earning power a given man had, the more social capital and gravitas he enjoyed among his peers and family. My informants were always quick to remind me that their parents or grandparents had 'come here with nothing', and that Britain provided a fertile ground to make money and 'get rich'. They would compare themselves with cousins in Pakistan or Bangladesh, concluding that they were fortunate to be in Britain, with all the economic opportunities that came with this. This memory of migration, and the will to work and provide, was a major component of what it meant to be a man. Here, my informants' 'style of masculinity' resembled what Osella and Osella refer to as the '*gulfan*' in Keralan communities with a history of high migration to the Persian Gulf. They argue that the newfound wealth and status of returnee migrant workers from the Gulf accentuated characteristics already locally associated with essentialised categories of masculinity.

> The *gulfan* [...] belongs to an intermediate category, not yet fully adult but with a central characteristic of adult maleness, money. Focus on cash as the defining characteristic of failed or successful *gulfans*, and the focus on consumer items brought and the expenditure while on visits at home, articulate with an idealized male life-cycle. Given that most *gulfans* begin their migration as young bachelors, leaving the village as immature youths (*payyanmar*), visits home are opportunities to demonstrate not only financial, but also age and gender-related progress. (Osella and Osella 2000:122)

Similarly, in Luton, young men conformed to essentialised or hegemonic notions of masculinity[6] that I suggest are relational to temporalities of migration and working-class socio-economic conditions. Being 'seen' as rich and, therefore, 'successful' was a preoccupation for many of my male informants. In some cases, this desire even led to careers in criminality. In fact, being a criminal was highly desirable for many of my younger interlocutors, who aspired to possess the power and prestige of locally known gangsters. There was a significant number of young men who left school without sufficient qualifications and entered a life of petty-criminality – usually dealing drugs or committing small-scale fraud. Once convicted, they found it near impossible to find legitimate employment. The worry for young Muslim men in Luton, therefore, is not that they are destined to become religious extremists, but whether they find a job before getting a criminal record.

The chapters of this book will provide a detailed ethnographic description of the everyday lives of my informants in Luton. I have organised the book into four core ethnographic chapters: 'Luton'; 'Family'; 'Friends'; and 'Religion'. The logic is to introduce the reader to the various arenas of socialisation and identity construction that young men in Luton are familiarised with from an early age and carry into adulthood. In doing so, my hope is to provide a holistic picture of concomitant expectations and pressures that young men continuously juggle in their everyday lives. The chapters also shed light on how my informants develop novel ways to manage, allay and rearticulate familial and social expectations and perceptions as a pioneering generation in their own right.

Chapter 1 introduces the reader to the town of Luton and, in particular, Bury Park where the largest concentration of the town's Muslims reside. It provides a geographical, social, and methodological context to the ensuing study. No other town in the British Isles with a significant Muslim population has received so much negative exposure since the declaration of the war on terror. Ever since the 7/7 bombers rendezvoused in the town prior to travelling on to London, Bury Park has been associated with a string of terrorism-related controversies. In 2009, a protest was held in Luton by a militant Islamist group objecting to the homecoming march of an army regiment returning from a tour in Afghanistan. This incident received considerable media coverage at the time and sparked off fresh debates pertaining to what constitutes acceptable protest among Muslim communities. In 2010, an Iraqi-born Swedish citizen detonated two bombs in the centre of Stockholm, Sweden, causing his own death and injuring two passers-by. This individual had spent much of his adult life residing with his wife and children in Luton, where he was perceived as a 'normal' member of the Muslim community. Luton is also the town where the radical far-right anti-Muslim organisation, the English Defence League (EDL) was founded in response to the above-mentioned protests held by Al-Muhajiroun in 2009. The EDL, originally named the 'United Peoples of Luton', is a street protest movement that opposes what it considers to be a spread of Islamism, Sharia Law and Islamic extremism in the UK. More recently, Luton has enhanced its reputation as a 'site of fear' with the brutal murder of Lee Rigby, a soldier in the British army, on the streets of London by two British Muslims claiming to act in the name of Islam. Rigby's murderers had been ideologically associated with some individuals from the town. In addition, a number of young British Muslims have recently travelled to Syria and Iraq to join Islamist militias in the region. With these wider contexts in mind, the chapter will seek to question perceptions of the town, and ask whether it really is a hot bed for political radicalisation and cultural segregation.

Chapter 2 moves on to shed light on my informants' relationships with family and the wider Muslim community. The vast majority of Luton's Muslims are of South Asian origin (mainly from Pakistan and Bangladesh), and are descendants of pioneer economic migrants that came to Britain from the New Commonwealth in the aftermath of the Second World War, in order to satisfy labour shortages at the time. Although some returned to South Asia once their financial needs were met, the vast majority opted to remain and permanently settle in the UK. Despite the fact that families had made the commitment to stay, the significance of maintaining links with South Asia and, in particular, cultural notions of kinship and community became very important. Community 'elders' were keen to enact South Asian etiquette, social relations, and moral/religious instruction within their households. This led to innovative re-enactments of South Asian family norms in a British setting. Whereas such norms were broadly accepted and expanded upon by many within the second generation – a significant proportion of whom moved to Britain as children or young adults – subsequent generations that have been born and raised in the UK were challenging the legitimacy of their South Asian heritage in providing a suitable cultural framework through which to live ostensibly 'British' lives. This chapter draws out some of these generational tensions and shows how young Muslims successfully manage expectations from the home alongside pressures from the wider society. The chapter demonstrates the way in which young Muslims accept and/or reject influences from both the 'South Asian home' and 'white liberal society' to create unique identities and subjectivities that succeed in undermining established notions of what it means to be British in the twenty-first century. The chapter charts the generational evolution of the community, bringing in voices from three generations of Luton's Muslims. It provides ethnographic insights into how early migrants settled in the town, their encounters with racism and exclusion, and the ways in which community solidarity helped them overcome those early challenges. These accounts are contrasted with how young Muslim men now manage to juggle the expectations from home with those from the outside world. This includes discussions pertaining to the ever-dwindling institution of arranged marriages, the development of intra-family gender relations, and expectations of moral conduct and 'honour' within family and community spaces.

Chapter 3 explores my informants' exposure to the world of work, making friends and having 'fun'. Even though Bury Park is a predominantly South Asian area of Luton, young Muslims were exposed to others from different ethnic and religious backgrounds from an early age. In addition to influences from the home, such interactions have profoundly shaped the way young Muslims self-identify and how they view others in society. In many cases, non-Muslims were regarded as close friends, mentors, colleagues, teammates, and even lovers. This chapter explores the role of social environments and spaces – such as school, university, sports and social clubs, and the workplace – in making friends and socialising. Muslim parents in Luton generally encouraged education, in line with broader trends within the wider South Asian diaspora. Consequently, a large number of young Muslims were pushed to regularly attend school and sixth-form colleges, and some went on to university. It is within these educational spaces that strong and lasting bonds were established with peers from all backgrounds. Similarly, those who did not continue with further education created relationships with neighbours and colleagues at work, were members of sports teams, attended youth centres, and visited the local mosque. Through these interactions, my informants were not only able to challenge and re-shape perceptions regarding their community, but also developed the ability to adopt and adapt to wider cultural and political paradigms. The chapter will aim to highlight how this process is *necessarily* integrative yet seldom remarked upon, and how it contributes to broader processes of multiculturalism that go beyond established state orthodoxies in policies and outlook.

Finally, Chapter 4 of the book investigates relationships and attitudes that young Muslims have developed with Islam and being Muslim in post-9/11 Britain. Since the declaration of the war on terror, British Muslim communities have come under considerable pressure to deal with the threat of radicalisation. Often, Muslim communities have been blamed by the state for the creation of closed, culturally segregated environments necessary for extremism to allegedly thrive. This has led to the enacting of numerous counter-terrorism laws and state-sponsored prevention strategies that disproportionally target British Muslims, causing community-wide resentment and suspicion towards the government. In addition, media depictions and discussions of Muslims as a 'fifth column' within society and a 'social problem' have cemented public discourses regarding Muslims that are recklessly reductive and homogenising. Curiously, in response to a wider atmosphere of hostility, young Muslims in Luton have seemingly gravitated more towards their religious identity as a means of expressing resistance to wider social perceptions, but also in solidarity with other Muslims both in Britain and around the world. For many young men in Luton, being a Muslim did not abrogate or undermine identifying with Britain, and they were keen to point out how they seamlessly merged. This chapter will seek to demonstrate how 'returning to Islam' has become an embedded symbol of resistance against Islamophobia, and how a distinct political identity plays out in the everyday lives of young Muslims. The chapter explores how notions of piety and spiritual devotion entangle with everyday political expressions of defiance against popular Islamophobia underpinned by responses to draconian legislation. It illustrates my informants' solidarity with fellow Muslims in other parts of the Islamic world, who are perceived as victims of aggressive Western foreign policies in the era of the 'war on terror'.

Notes

1. See Abu Lughod 2002; Ward 2006; Jouili 2009; and Fadil 2009.
2. Compare with Alexander 2000; Archer 2003; Alam 2006; and Hopkins 2008.
3. See Alexander 1996.
4. Compare with Archer 2003.
5. See Morey and Yaqin 2011.
6. See Archer 2003.

Discussion

This book is a contribution to scholarly approaches to the anthropology of Islam, both past and present. Anthropological approaches to the study of Islam and Muslims have been much contested in the sub-discipline's relatively short history. Initial interest was sparked by Clifford Geertz's (1971) famous thesis on the importance of religious symbols in providing cosmological meanings to individuals' everyday lives. This 'cultural' view was challenged by Ernest Gellner's (1981) sociological interventions, which argued that Islam is a means by which society is ordered. Gellner suggests that, since Islam is a scriptural religion that lays down permanent laws and principles pertaining to how to organise society, and given that that scripture was 'completed' by the mission of the Prophet Mohammed, Muslim society is one where religion and the state become enmeshed as one. Sociological order, therefore, is sustained by a class of proto-politicians who are also religious authorities at the same time, much like the Prophet himself. Further, it is this class of scholar-cum-politician that provides for social order and consequent solidarity in Islamic communities. Michael Gilsenan (1982), however, took a different approach. He argued that Islam or religion should not necessarily be the primary focus for anthropologists when studying Muslim societies. Rather, other social, economic and political factors need also be incorporated if one is to attain a holistic ethnographic account of the people and communities under study. For Gilsenan, then, Islam was indeed a core component of life in the Middle East. However the weight of other conflicting and parallel social phenomena co-existing with 'the religious' meant that any account where religious symbols or agents provide a totalising coherence to quotidian lives did not suffice as an accurate ethnographic picture.

In tandem with these epistemological developments, was the idea proposed by Abdel Hamid El-Zein (1977) that there is no such thing as an 'anthropology of Islam'. His work implies that both Geertz and Gellner were guilty of 'essentialising' Islam in their respective symbolic and sociological approaches. For El-Zein, 'Islam' was only relevant in local contexts where the religion took particular vernacularised and idiosyncratic social forms. How Islam was socially articulated in Tanzania is quite different from, say, Indonesia, for example. The logical conclusion, therefore, is that ethnographic evidence suggests that there is no *one Islam*, but a collection of *Islams* that exist and have always existed in various parts of the world. It is thus the task of the anthropologist to draw out these differences for the purpose of comparative analysis which, he posits, is the intellectual basis of the discipline. Perhaps the most ardent critic of this perspective is Talal Asad (1986). In his seminal essay 'The Idea of an Anthropology of Islam', Asad attempts to identify and define Islam as an 'analytical object of study'. He argues against his forebears that Islam can be reduced to a system of symbols and social structure, or that it is a 'heterogeneous bundle of beliefs, artefacts, customs and morals' (1986:14). Instead, Asad claims that Islam is a 'discursive tradition' which is ultimately based on reference to the Qur'an and Hadith (biographical accounts of the Prophet). He suggests that all Muslims defer to the holy texts of Islam, and place them above all other forms of ontological reference. In order to conduct a 'proper' anthropological study of Islam, therefore, one must be prepared to locate particular religious or theological discourses that inform Muslim practitioners. Moreover, Asad alerts us to the idea that all Muslims seek and strive for a sense of religious 'orthodoxy' in their lives, regardless of the respective sectarian differences that exist between different Muslim groups. All Muslims, he claims, are invested in practising Islam in ways they deem to be acceptable in the eyes of God, linked to particular conceptions of temporality. This 'discourse' is always associated with scripture but manifests in different ways in different social contexts:

> Orthodoxy is crucial to all Islamic traditions [...] [It] is not a mere body of opinion but a distinctive relationship – a relationship of power. Wherever Muslims have the power to regulate, uphold, require, or adjust *correct* practices, and to condemn, exclude, undermine, or replace *incorrect* ones, there is the domain of orthodoxy. The way these powers are exercised, the conditions that make them possible (social, political, economic etc.), and the resistances they encounter (from Muslims and non-Muslims) are equally the concern of an anthropology of Islam, regardless of whether its direct object of research is in the city or in the countryside, in the present or in the past. Argument and conflict over the form and significance of practises are therefore a natural part of any Islamic tradition. (Asad 1986:15 original emphasis)

Discerning 'correct' practices from 'incorrect' ones, therefore, is the vital part of any anthropological toolkit when exploring the lives of Muslims, according to Asad. Any endeavour that does not begin from the concept of a discursive tradition relating to the founding texts of Islam – 'as Muslims do' – is futile since Islam is, first and foremost, 'a tradition'.

Asad's sophisticated and seemingly radical contention was well-received by some, and a number of notable works have since emerged where Islam is indeed understood as a discursive tradition.[1] These works broadly trace the role of religious law in informing ethical practices in the lives of 'pious' Muslims. Although they do not deny the existence of a world beyond the mosque or religious court per se, protagonists are usually located within a habitus where religious seeking for the sake of spiritual coherence is pronounced. This can be misleading, since not all Muslims are seeking ethical and spiritual perfection all of the time (if this is indeed possible). Moreover, a number of more recent ethnographic accounts have demonstrated that Muslims are, more often than not, constantly oscillating between and traversing through a number of concomitant 'moral registers', 'regimes of truth' and 'multiple identities'.[2] These scholars propose that, instead of focusing on textual Islam as an informative source for conducting ethnography, a more nuanced approach would be to demonstrate how Islam is 'lived' by its adherents. Through this method, anthropologists can observe the points of disjuncture, ambivalence, and everyday slippages that litter the lives of Asad's 'discursive' Muslim searching for orthodoxy. At the same time, the lives of 'non-practising', 'lapsed' or 'nominal' Muslims can also be absorbed into the anthropological gaze. In so doing, a more beleaguered, torn and incoherent subject is brought to the fore, thereby demonstrating the complexities and contradictions of Muslims lives. In fact, Samuli Schielke (2010), a keen advocate of the 'lived Islam' approach, has even suggested that 'there is too much Islam in the anthropology of Islam'.

Although there is much to take note of from Schielke's intersectional insights as a methodological starting point – particularly with regards to moving focus away from the 'ethical habitus'[3] – my own exposure to Muslims in Luton suggests 'Islam', 'the text', and a quest for 'orthodoxy' maintained resonance with all of my informants (I did not meet anyone who claimed not to believe in Islam). Moreover, 'being Muslim' not only meant striving for moral coherence (however difficult and thankless), but brought with it a distinct political identity that, I suggest, is particular to and seemingly acute among Muslims living in the West.[4] The vast majority of my informants struggled with being a 'good Muslim',[5] but this did not deter them. Moreover, being Muslim in the post-9/11 world engendered my informants with a will to 'represent' and 'defend' their religion against a perceived tide of antagonism and suspicion from White liberal society. I spent time with young men who purposefully missed prayers, refused to fast, dealt drugs, fornicated regularly and yet argued that women should wear the *hijab*, planned trips to Mecca for pilgrimage and always ate *halal* food. This 'ambivalence', as Schielke rightly puts it, stood side-by-side with a cogent, conscious and coherent sense of being 'Other' within their own society. This manifested most prominently in a suspicion towards the discursive and institutional influences of the state,[6] and a longing to be contingent with the utopian habitus of the international fraternity of Muslims (*ummah*).[7]

In the post-9/11 world, being a young Muslim in diaspora offers up a particular set of challenges. There is a qualitative difference between my informants and their peers living in Muslim-majority countries.[8] It is premised on a particular history and retrospective memory of migration that is intrinsically linked with the global flow of colonial labour in the aftermath of the Second World War.[9] These economic 'push factors',[10] coupled with subsequent immigration restrictions[11] led, eventually, to the permanent settlement of mostly Caribbean and South Asian families in Britain. Of this cohort, Muslims comprised the largest ethno-religious group (predominantly originating from Kashmir, Punjab, Gujarat and East Bengal).[12] My young informants in Luton were well acquainted with stories of such migration, the struggles that pioneer migrants faced when they first arrived and, importantly, how the fortunes of their family (both in Britain and South Asia) were revolutionised as a result.

Once settled in Britain, these working-class 'strangers' were confronted with a host of institutional and political hurdles, particularly in the struggle for ethnic and racial recognition.[13] Despite a sustained period of unified race-based political mobilisation in the 1970s (the decade when immigration from the New Commonwealth was permanently curtailed), previous unity became fragmented following the advent of Multiculturalism as state policy in Britain. This was implemented in reaction to a series of race riots in the early 1980s, staged within ethnic 'ghettos' across the country.[14] The British government's response to structural racism, (on this occasion, in the specific form of police brutality), was to instigate a set of policies that sought to 'celebrate cultural diversity'.[15] In effect, it disintegrated previous class solidarities, as various 'ethnic' groups jostled for newly ring-fenced state funds and recognition. This period of structural change, I suggest, produced certain material and discursive conditions that succeeded in cementing particular conceptions of the British Muslim community (both internal and external) that survive to the present day. Moreover, arguably no other event (including 9/11) has played a more significant role in simultaneously illuminating and demonising British Muslims than the 'Satanic Verses' affair of 1989.[16] Multiculturalist policies implemented throughout the 1980s attempted to institutionalise essentialist conceptions of minority 'cultures' by encouraging different minority groups to vie for government subsidies: 'Different ethnic groups were [thus] pressed into competing for grants for their areas. The result was that black communities became fragmented, horizontally by ethnicity, vertically by class' (Kundnani 2002:69).

British Muslims, being one such group, were propelled into the public imaginary by the fervent protests and public burning of *The Satanic Verses* in Bradford, a northern ex-mill town where a substantial Muslim community continues to reside. These scenes were received with horror by White liberal society, as sensational images of people burning books, circulated in the press, brought back memories of Nazi Germany.[17] Prior to this watershed moment, British Muslims were seen as 'Asians' or even as politically 'Black'.[18] The *Satanic Verses* affair separated British Muslims from other minority groups. It made them 'visible'. Moreover, it was a public expression of the Multiculturalist vision, albeit an unintended one. Muslims became identified along ethno-cultural lines, along with other minority groups, such as British Sikhs for example.[19] Culturally, they were no longer 'British-Pakistanis' but were now primarily recognised as 'British Muslims'. The controversy created a seminal shift in both how the wider British public viewed Muslims within their midst, and also how Muslims themselves self-identified. By the time my 'millennial' informants witnessed the events of 9/11 and, later, the London bombings of 2005 (7/7), they were already socialised along ethno-religious lines. Multiculturalist policies and public discourse, therefore, played a direct role in cultivating communal identities. Notably, the implementation of such policies followed on from the breakdown of trust between ethnic minority groups and the state, caused by fundamental structural inequalities. Moreover, although similar patterns of settlement could be observed throughout Europe and North America at the time,[20] the British case is specifically contingent on a particular colonial genealogy.[21] Any credible ethnographic inquiry into British Muslims of South Asian origin, therefore, must be sensitive to these formative historical machinations.

Such conditions, specific to diaspora, provide an additional metonymic node to the anthropological study of Islam in order to move beyond classical fixations on tribes, harems and 'men of piety'.[22] I suggest that its catchment must include the study of diaspora, post-colonial subjectivities and transnational social forms. This book thus attempts to supplement anthropological works situated in the 'centre' of the Islamic world, through an appraisal of its 'peripheries'. It draws ethnographic attention to how state narratives and policies possess the potential to balkanise its citizenry, explores the teleological impact of migration and generational renewal, and demonstrates how Muslims born into minority status are developing novel political identities in the post-9/11 world. In doing so, I hope to expand on classical and contemporary debates within the anthropology of Islam, and provide an area of inquiry that embraces *both* textual and quotidian approaches to studying Muslims. Moreover, an inquiry that sheds ethnographic light on the lives of those Muslims who, as Olivier Roy (2004) argues, are 'culturally' and 'territorially' detached from the Islamic world.[23]

***

By the time the events of 9/11 became known, Multiculturalist doctrines were fully embedded. As the fallout from the atrocities unfolded, Muslims in Britain already shared a common identity that was, as we've seen, endorsed by the state. The months following on from those events were a particularly tense and anxious time for British Muslims. Many faced various forms of discrimination, from bearded men being refused entry onto public coaches, to the overnight termination of large commercial contracts with Muslim-owned businesses, to veiled women being abused in the streets.[24] All of my informants retained significant memories of this period, and many of them held strong political views on the subsequent 'war on terror'. It is also important to note here that many of my informants came of age in a political milieu dominated by events concerning Muslims from around the world. I have already discussed the significance of the Rushdie affair in 1989 in pushing the British Muslim community into the public gaze. The Rushdie affair was shortly followed by the outbreak of wars and violent skirmishes involving civilians in a number of Muslim-dominated regions in Europe (Bosnia Herzegovina, Kosovo), and Asia (Chechnya, Palestine, Kashmir). Notably, they all took place prior to 9/11. Images and reports of destroyed mosques, widespread displacement and carnage, and high civilian death tolls were consumed by my interlocutors on a regular basis. British Muslims, who considered themselves as part of the international Muslim community (*ummah*), were active in campaigning in support of the victims, consistently lobbying the government to reach resolutions. So many of my informants (particularly those old enough to remember) claimed that they felt the same sense of solidarity with Muslims from other parts of the world as they did with fellow British Muslims.[25] Moreover, this is precisely the arena where the emotional efficacy of a common 'discursive tradition' as Asad describes it, can be observed.

The idea of the *ummah* is one sanctioned by the holy scriptures in Islam. Specific nuances of its meaning and definition have often been disputed by theologians. However, the notion of a common humanity and solidarity between Muslims is one supported by unequivocal references in the Qur'an and Hadith literature:

> Indeed, [all] Muslims are brothers (Qur'an: Al-Hujarat: 10).
> The example of Muslims in their mutual love, mercy and sympathy is like that of a body; if one of the organs is afflicted, the whole body responds with sleeplessness and fever (Hadith-Muslim).

These references were often memorised and recited by my interlocutors to demonstrate the encompassing Islamic obligation to stand united with co-believers across the world. Such transnational loyalties almost always contradicted British foreign policy throughout the 2000s, (including during my own fieldwork). Particularly pertinent for our purposes here, however, is the almost universal consensus among my informants, young and old, that Western foreign policy in Iraq, Afghanistan and Palestine was unacceptable and tantamount to a 'war against Muslims'. Many claimed the only motivation for military interventions in Muslim lands was due to a scramble for natural resources. My interlocutors were thus highly critical of the British state, with their loyalties clearly aligned to the fate of co-religionists rather than the interests of their country of residence. Terence Turner posits such divided loyalties as the 'postmodernist reaction to the delegitimization of the state and erosion of the hegemony of the dominant culture in advanced capitalist countries' (1993:423). He suggests that:

> [P]eople all over the world have turned to ethnic and cultural identity as a means of mobilizing themselves for the defence of their social and political-economic interests. The increasing political importance of culture as an ideological vehicle for the new forms of ethnic nationalism and identity politics that have accompanied the weakening or collapse of colonial empires and multiethnic states, has made it a favoured idiom of political mobilization for resistance against central political authorities and hegemonic national cultures. (Turner 1993)

My informants in Luton certainly maintained an unmistakable camaraderie with those they perceived to share a common cultural and historical heritage. Many attended and organised marches, rallies and lectures in solidarity with Muslims in Palestine and Iraq, for example. Unity with international Muslims was complemented by unity with those living in the UK, who were also included within the aegis of the *ummah*.[26] Here, too, fellow Muslims were perceived to be victims of the 'war on terror', most pronounced through the manner in which Muslims had been profiled in the media;[27] and, perhaps more significantly, in reaction to the government's PREVENT strategy, designed to tackle extremism and radicalisation. Studies have shown that British Muslim communities have been largely suspicious of the latter[28]. Counter-terrorism policies are grounded in 'intervention strategies' that theologically challenge 'extremist' positions. Such strategies rely on the cooperation of Muslim communities in identifying and participating in interventions, while also paradoxically targeting them as a 'suspect community'[29]. Being seen as a target community and, at the same time, being expected to lead the charge against radicalisation has led to widespread confusion and resentment among Muslims and mistrust towards government. Muslims felt compelled to 'apologise' for their religion as a result of the actions of militants around the world who purport ideas and actions that many consider to be acts of overt political transgression rather than any attempt at scriptural conformity.[30] In the face of the perceived international persecution of Muslims as a result of Western foreign policy, and the further marginalisation of Muslims at home, depicted as a 'suspect community' or 'fifth column', the conception of the *ummah* as an emancipatory social form and 'site for resistance' took root among my young informants. Feeling for the *ummah* was not only a religious obligation, but also a political necessity for the times. By being conscious of and, in many cases, overtly expressing resentment and criticism of British foreign and domestic policy towards Muslims, my informants were actively in the process of subverting established state narratives of citizenship and belonging. Moreover, my informants were able to navigate this 'double consciousness', as W.E.B. Du Bois aptly put it, in one of two ways. They had recourse either to an abstract, imagined utopia of the *ummah* in the face of wider hostilities, or to what Franz Fanon claimed to be 'some very beautiful and splendid era whose existence rehabilitates us both in regard to ourselves and in regard to others' in order to escape 'the misery of everyday life, self-contempt, resignation and abjuration' (Fanon in Hall 1993:394).

The *ummah* thus acted as an abstract 'space', that Gupta and Ferguson (1997) describe as 'a neutral grid' within which post-modern transnational communities inscribe 'cultural difference, historical memory and societal organisation' (1997:34). However, *ummah*, as my informants understood it, further subverts this idea to necessarily create a space that is *devoid* of culture, history, or society in order to suit the culturally flattening conditions of diaspora. Moreover, my informants *politicised* its meaning. It was a utopian habitus where categories of class, race and gender collapsed. The *ummah* was an inclusive realm within which all Muslims of the world were united. The way in which Muslims in Luton conceptualised the *ummah* forms a fundamental component of my argument in this book. I argue that young Muslims in Luton are increasingly becoming attracted to what Olivier Roy refers to as globalised Islam (2004). He suggests that Muslims living in the West resist culturally alienating state narratives, as well as the equally alien South Asian 'culture' of their parents, by valorising their Muslim identity, which transcends national or ethnic allegiances. Roy argues that the combination of globalisation, westernisation and the increase in worldwide Muslim diasporas has led to a reimagining of the *ummah*. A 'global Muslim' can mean either:

> Muslims who settled permanently in non-Muslim countries (mainly in the West), or Muslims who try to distance themselves from a given Muslim culture and to stress their belonging to a universal *ummah*, whether in a purely quietist way or through political action. (2004:ix)

Roy claims that second and third generations in diaspora are particularly attracted to the doctrines of globalised Islam due to the fact that they are 'de-territorialised' from the Islamic heartlands. Moreover, they are dissuaded by the alienating cultural proclivities of both the host community, and particular non-Islamic cultural preoccupations of preceding generations. Instead, the appeal of being a member of the *ummah*, united by a common belief that transcends race, ethnicity and nationality is particularly powerful. In the post-colonial world, the '*ummah* no longer has anything to do with a territorial entity', but now must be 'thought of in *abstract* or *imaginary* terms' (2004:19 my emphasis).

He goes on to suggest that Muslims in diaspora are also captivated by the notion of a 'glorious Islamic past' promoted by various reformist thinkers and groups. This conception of history resonated with many of my interlocutors in Luton, especially those disenfranchised by the regional 'folk Islam' professed by members of the older generations.[31] Young Muslims regarded the Islam practised in the home to be more 'cultural' than 'religious'.[32] They argued that their parents often conflated

South Asian 'cultural' practices with Islam. This realisation led many to engage in debate and discussion with their parents, mosque *imams* and elders in the community.[33] The pursuit for a 'pure' Islam, devoid of polluting cultural influences, is consistent with what Roy's defines as 'deculturalised Islam':

> The construction of a 'deculturalised' Islam is a means of experiencing a religious identity that is not linked to a given culture and can therefore fit with every culture, or, more precisely, could be defined beyond the very notion of culture [...] The new generation of educated, Western born-again Muslims do not want to be Pakistanis or Turks; they want to be Muslims first. (2004:22–5)

Homi Bhabha (1994) suggests that explorations of the fragmented cultural identities of post-colonial subjects, particularly those generated within the modern diaspora, must escape essentialist paradigms and seek to unearth the discursive spaces that exist 'in-between' dominant political and ontological discourses. Bhabha's proposition of the 'third-dimension' is particularly pertinent for our purposes here. He argues that the historical ruptures and transformations which occur as a result of mass movement of people across continents profoundly disrupt the applicability and generation of previously held 'traditions'. Confronted with a new environment, minority subjectivities embark upon a process of novel and innovative re-articulation. He claims that:

> What is theoretically imperative, and politically crucial, is the need to think beyond narratives of originary and initial subjectivities and focus on those moments or processes that are produced in the articulation of cultural differences. The "in-between" spaces provide the terrain for elaborating strategies of selfhood – singular or communal – that initiate new signs of identity, an innovative site of collaboration, and contestation, in the act of defining society itself. (1994:2)

Bhabha focuses on the idiosyncratic character of new 'hybrid' forms of identity that occur as a consequence of cultural interaction. He identifies a 'third-space', which is 'invisible' to prevalent homogenising discourses of the nation-state, where cultural meanings and subjectivities are continually reproduced and innovated to create cosmopolitanisms that transcend and undermine national narratives of cultural homogeneity. Bhabha's approach corresponds with the way my own informants generated new identities in the 'in-between' spaces of the home and wider society, informed by the attitudes and discourses that dominate both spheres, and yet are distinct from it.

Diaspora cultural (re)production of this kind is further verified by the works of Paul Gilroy. Based on research conducted among Britain's post-war Afro-Caribbean communities, Gilroy locates the production of black identities in the diaspora within the milieu of anti-racism struggles in Britain throughout the 1970s and early 1980s. He focuses on the popular appeal of politically subversive Afro-Caribbean music produced in this period. Through various cultural mediums, Gilroy contends that black youths developed dynamic forms of political resistance that were unique and flexible and that went beyond classical forms of proletarian organisation.[34] He is critical of traditional Marxist cultural approaches that seek to simply situate working class 'black expressive cultures' as characteristically 'anti-racism'. He suggests that Black-British culture transcends the narrow confines of the politics of anti-racism and ethnic absolutism and incorporates all the taxonomical complexities of 'race', 'ethnicity', 'nation' and 'culture', as well as 'class', to produce something new and unremarked. He writes:

> Black Britain defines itself crucially as part of a diaspora. Its unique cultures draw inspiration from those developed by black populations elsewhere. In particular, the culture and politics of black America and the Caribbean have become raw materials for creative processes which redefine what it means to be black, adapting it to distinctively British experiences and meanings. Black culture is actively made and remade. (1987:202)

Gilroy alludes to a process of 'cultural syncretism', which emerged within the black diaspora in Britain that had galvanised at a particular political moment. Yet, at the same time, he reminds us that the diaspora should not be analysed as separate from the British social fabric. Black culture, he argues, should be considered a significant interlocutor within the 'ever-in-process' British cultural mosaic:

> An intricate web of cultural and political connections binds blacks here to blacks elsewhere. At the same time, they are linked in the social relations of this country. Both dimensions have to be examined and the contradictions and continuities which exist between them must be brought out. (1987:205)

This approach is entirely apposite within the context of my interlocutors in Luton. Like black youths in the 1980s, young Muslims have developed politicised identities in the face of turbulent global events and the burgeoning of local suspicions against them. In contrast to Gilroy's black youth, it is not necessarily their 'race' which is divisive in social relations, but their religion. Moreover, it is the dual cultural and political capacity of the imagined *ummah* that 'binds [Muslims] here to [Muslims] elsewhere' within a domain of mutual struggle and solidarity against neo-liberal forces of domination.

Similarly, Stuart Hall reminds us that identity should not be viewed as an already accomplished fact, but as a '*production* which is never complete, always in process, and always constituted within, not outside, representation' (Hall 1993:392 my emphasis). He posits that displaced communities attempt to collectively essentialise identity in an attempt to impose a 'shared culture', even where this process seems difficult to engender.[35] Cohesion, he argues, is achieved through displaying unified projections of ideological solidarity even where such solidarity is superficial or artificially imposed. Borrowing from Fanon, he suggests that the experience of colonisation has contributed towards the 'emptying of the natives' brain of all form and content'. Colonial authorities, he posits, reshaped the historical trajectories of the subject populations to the extent of 'distorting, disfiguring and destroying them' (1993). Once fully conscious of these historical mutations, the colonised attempt to unearth that which the colonial experience has buried and overlaid. He concludes that:

> Cultural identities are the points of identifications, the unstable points of identification, which are made, within the discourses of history and culture … not an essence but a positioning … hence there is always a politics of identity, a politics of position, which has no absolute guarantee in an unproblematic, transcendental law of origin. (Hall 1993:395)

Black-British identity has emerged from heterogeneous roots as slaves were taken from different countries, tribal communities, villages, linguistic groups and religious traditions. However, the specific conditions of displacement and diaspora have led to the generation of novel Black-British subjectivities that seek to reach beyond the confines of the exclusivist nation state. Black-Britishness functions as a unifying counter-culture against the trials of social marginalisation. Similarly, in Luton, the *ummah* was conceptualised as an effective means of reconciling a sense of perceived 'Otherness' in society, made possible by Orientalist stereotyping of Muslims as exotic, visceral and violent in the popular imagination.[36]

My informants, self-identifying as they did, must thus be contextualised within specific historical, social, economic and political developments that have shaped their intersectional and often contradictory opinions, emotions, and loyalties throughout their lifetime. These conditions have given rise to unique vernacular and hybrid forms of British Muslim identity based on national and transnational solidarities, personal piety and salvation, and a recalibration of what it means to be British in today's post-colonial Britain.[37]

Notes

1. Compare with Mahmood 2001; Hirschkind 2001; Henkel 2005; Deeb 2006; and Bowen 2012.
2. Compare with Marsden 2005; Soares 2005; Simpson 2008; and Schielke 2009.
3. See Mahmood 2001.
4. See Roy 2004; and Fernando 2016.
5. Compare with Simon 2009; and Schielke 2009.
6. Compare with Herzfeld 1985.
7. Compare with Stewart 2017; Mandaville 2004; and Roy 2004.
8. See Masquelier and Soares 2016.
9. See Castles 1987; Solomos 1993; Ballard 1994; and Ansari 2004. Compare with Fanon 1967.
10. See Kapoor 2002.
11. See Patterson 1969:17–34.
12. See Clarke *et al* 1990.
13. See Castles and Koszak 1973; and Phizaklea and Miles 1980.
14. See Kundnani 2002; compare with Parekh 2000; and Modood 2007.
15. See The Scarman Report 1981; and Benyon 1984.
16. *The Satanic Verses* is a novel written by British author Salman Rushdie, which was published in 1988. The book attracted controversy as some Muslims claimed it was a deliberate parody of the life of the Prophet Mohammed, and therefore deemed to be blasphemous. The affair was heightened by the issuing of a *fatwa* (religious decree) by Ayatollah Khomeini, then the Supreme Leader of the Islamic Republic of Iran. The decree urged all Muslims to target Rushdie for defaming the name of the Prophet, and to seek to assassinate him. In the UK, Muslims reacted to the *fatwa* by a staged public book-burning in Bradford. This event attracted considerable media attention, and propelled British Muslims into the public eye as separate from the wider South Asian diaspora for the first time.
17. See Parekh 2000:295–335; and Modood 1990.
18. See Shukra 1998.
19. See Bhachu 1985; Clarke *et al* 1990; Ballard 1994; Byron 1994; and Hall 2002.
20. See Mandel 2008; Silverstein 2004; and Wikan 2002.
21. See Ansari 2004; Visram 2002; and Gilroy 2006.
22. See Abu-Lughod 1989.
23. Compare with Appadurai 1996; Linke 2014; Turner 1993; Grillo 2003; Johnson 2012; and Vertovec 1997.
24. See Sheridan 2002: 91–92; and Bunglawala 2002.
25. Compare with Lewis 1994:164–9; Werbner 2002:153–183; and Jacobson 1998:149–151.
26. Compare with Archer 2009.
27. See Poole 2002; Meer 2006; Allen 2010; and Morey and Yaqin 2011.
28. See Spalek and McDonald 2010; Heath-Kelly 2013; Hopkins and Gale 2009; Githens-Mazer 2012; Lister and Jarvis 2013; and Kundnani 2012.
29. See Awan 2012; and Breen-Smyth 2014.
30. See Choudhury and Fenwick 2011; and Hopkins 2009.
31. Compare with Robinson 1983; Das 1984; and Van der Veer 1992.

32. Compare with Lewis 1994:197; Jacobson 1998; and Kibria 2011.
33. Compare with Marsden 2005.
34. See Gilroy 1987.
35. Compare with Grillo 2003.
36. Compare with Said 1978; Lewis 1990; and Huntington 1996.
37. Compare with Alexander *et al* 2003; Modood and Werbner 1997; and Appadurai 1995.

1
Luton

Luton is a horrible town. I commuted there first from my home in Surrey and later from my new residence in East London. Sometimes I drove but usually I took the train. My field site was mainly centred around Bury Park, an area close to Luton's town centre. Bury Park is a very interesting place. There the majority of Luton's share of post-war New Commonwealth migrants live and, in particular, it is the home of Luton's substantial Muslim community.[1] The majority hail from Muslim-majority parts of South Asia, particularly from Pakistan and Bangladesh.[2] I did not know it at the time of research but Luton is also home to the English Defence League (EDL). The EDL is a far-right protest movement with branches all over the British Isles. Although they claim not to be racist, its supporters are unapologetically Islamophobic.[3] Luton itself has often been criticised by the EDL (as well as some outside observers) as being a hub for religious (read: *Islamic*) radicalisation. This is thought to be taking place within its Muslim community. 'Militant Islam' as the former leader of the EDL, Tommy Robinson, described it. Robinson (real name Stephen Lennon) grew up in Luton, but not in Bury Park. He lives instead in a predominantly white working-class part of the town. He does not often venture into Bury Park because, in his own words, it's a 'Muslim part of town'. *His* town. Years of publicly ranting against Muslims and their religion has made him a marked man on the other side of town. Being a pariah in his own town is said to be a driving force behind his activism. That and the endless attention he receives from the British mainstream media.

The straw that finally broke Robinson's back, however, came in 2009, a few months after I had completed fieldwork. Luton was set to host the homecoming march of the Royal Anglian Regiment, returning from a tour in Afghanistan. It was to face resistance. A dozen or so members of the hard-line Salafi[4] group Al-Muhajiroun (AM) – mainly from outside of Luton – organised a counter-rally. They heckled the troops, shouting out 'Terrorists' and 'Butchers of Basra' as the returning veterans marched past. They also held up placards with 'Shariah for the UK' and 'Hands Off Muslims'. Back then, AM believed in establishing a worldwide 'Islamic' caliphate.[5] Their ultra-literalist creed (and political ideology) dictates that establishing *Shariah* rule is a fundamental tenet of faith (*aqeeda*), and Muslims everywhere and anywhere should be steadfast in the fulfilment of this duty (*fard*). Naturally, of course, for AM this call also applied to Muslims living as a minority in the UK. Interestingly, at the time, AM only operated in the UK. It had no branches, influence or followers (for that matter), in any majority-Muslim country. In response, Robinson and his friends created the United Peoples of Luton (UPL). The UPL later morphed into the nationwide EDL as it grew in momentum and consolidated its identity as an anti-Islam nationalist movement.

Bury Park is also infamous for being the home of the so-called 'Stockholm bomber', Taimor Abdulwahab al-Abdaly – an Iraqi-born Swedish citizen who detonated a homemade bomb in the centre of Stockholm in 2010, killing himself and injuring two passers-by. He was said to have lived a 'normal life' in Luton with his wife and small children. Abdaly attended the same Salafi mosque where I conducted fieldwork, and was known as a quiet but 'decent' family man by members of the congregation that knew him. Furthermore, Michael Adebolajo and Michael Adebowale, the two men that killed the British Army soldier Lee Rigby in 2013, were also identified as having strong links with Luton. As members of AM, the two regularly appeared in the town for missionary (*dawa*) purposes, and on many occasions tried to recruit some of my informants. Since these events, ISIS established its version of the caliphate in parts of Iraq and Syria. Some AM members consequently left Britain to join the movement. AM's leader, Anjem Choudary, was jailed in 2016 for supporting ISIS. He too was a regular visitor to Luton, delivering public speeches and campaigning for the caliphate in the streets of Bury Park. In 2017, Khurram Shazad Butt, a member of AM, was responsible for the London Bridge attacks, killing eight members of the public. Most recently, Khalid Masood, the 2017 Westminster Bridge attacker, who killed five people and injured over 50 more, was also a one-time resident of Luton. In fact, he was employed as an Arabic teacher at the very mosque where I had conducted fieldwork. His views and activities at the time, however, were unknown to most in the congregation.

Luton, then, has seen some very unsavoury characters. It has often been perceived as a 'site of fear' due to its links to terrorism and various forms of radical political groups. It is also a post-industrial landscape and, as such, it is a bit of an eyesore. Not least around Bury Park. The centre of town – where I got off the train – is located about a ten-minute walk away. It has a rather unremarkable pedestrianised high street containing the usual chains of retail outlets and banks. Bury Park is located to the north-west of the centre and, to get to it, one has to negotiate a gauntlet of grubby steel and concrete in the shape of ring-roads, roundabouts and urine-infested subways. A sharp turn and the landscape changes, as well as the decibel levels. Bury Park itself is home to row upon row of poky red-brick terraced houses from a bygone era. The tiny 'gardens' behind the front walls were usually jammed full of an assortment of green and purple 'wheelie' bins, and the occasional inflatable crocodile. The faces are also different from those in the centre of town. The majority of the residents here are South Asian in origin and Muslim in religion. This is most tellingly illustrated by the assortment of mosques that I pass by. Luton has a total of twenty-five mosques, the majority of which are located in Bury Park. Mosque congregations here are dependent on sectarian orientation, ancestry and pragmatism. I pass a mosque attended predominantly by the town's Konkani Muslim Community,[6] another aligned to Saudi Arabia, and another with a majority Bangladeshi-origin congregation.

Eventually my path converges with Dunstable Road, the busiest section of which is a semi-pedestrianised retail district, stretching for approximately a quarter of a mile. This is where it all happens. The thoroughfare is home to dozens of saree shops, halal butchers, fast-food outlets, restaurants and grocery stores – the majority of which are owned and staffed by people who are ostensibly of South Asian origin. Dunstable Road is always busy with people shopping, eating, catching up with friends, and even preaching. Once a week, the chaos calms for an hour or so, as hundreds of Bury Park's Muslims jostle past each other in different directions towards their preferred mosque in order to observe the Friday (*Juma'a*) prayers. This colour and vibrancy does well to mask the stark and quite visible social and economic deprivation of the area. After a while (and a curry or two) it does not seem so horrible.

***

Dunstable Road is surrounded by a large residential area on either side. Except for the main road arteries, the minor roads are all punctuated by speed humps in order to reduce questionable driving. A good number of locals, however, have cleverly by-passed this problem by owning large off-road vehicles. These vehicles can not only drive straight over the humps as if they were not there, but also serve as status symbols for their owners, especially for the older generations. The younger generations, however – mostly young men – have developed a preference for customised sports cars. These cars have lower, more rigid suspension systems that require time and delicacy to negotiate the unwelcome obstacles. The residential roads often become quite congested with a trail of cars, to the sometimes vocal fury of the 4x4 owners, clearly upset that their shrewd investments were failing to yield.

This neighbourhood is majority-Muslim, and the terraced houses here consist of three or four bedrooms. At the time of fieldwork, there was a popular trend of extending houses in order to accommodate burgeoning extended families. Households are known to compete with each other over the quality, size and aesthetics of such extensions. In addition to the cramped front garden, the back gardens where small children will often play during the summer months are also quite compact. Inside, the rooms are small and space was often congested due to the number of people living there. Households consisted of three or more generations. Parents would usually live with their children until they either got married and left (in the case of daughters), or there was no more room left due to the number of grandchildren (in the case of sons). In the latter case, a nuclear group would either detach from the main household and move nearby, or the entire household would relocate to a larger house, if they could afford it. The latter option, however, usually meant moving out of Bury Park; an option to which many of my informants were resistant. As a consequence, Bury Park is quite distinct as far as residential areas in Luton are concerned. It is over-populated and intergenerational – a collection of large, extended British-Asian households living side-by-side in small, often eccentric, terraced and semi-detached houses. This part of town, however, shares marked aesthetic, social, economic and cultural similarities with a number of other areas of the UK with a high density of post-war South Asian economic migrants from the New Commonwealth.[7]

***

The presence of Muslims in Luton extends as far back as the 1960s, consistent with the mass migration of South Asian and Caribbean workers in that period.[8] Initially, mainly male migrant workers came to Luton in order to seek employment in the then flourishing manufacturing industries situated in and around the town. Luton is thirty-two miles north of London, and is served by major intercity rail networks. It is also located at Junction 9 of the M1 motorway, which was built in 1959 as a major highway connecting London and Leeds. Luton also boasts an international airport, serving London and the south of England. The town's close proximity to the capital, and its rich access to major road, rail and air networks, along with cheap land and renting tariffs, facilitated the establishment of a substantial manufacturing sector throughout the twentieth century. According to my informants, many of the pioneer migrants first sought work at the Vauxhall Motors plant. Established in 1905, Vauxhall Motors employed around 30,000 workers from the neighbouring areas at its height.[9] It was not only new migrants who flocked as workers to Luton, but also many settled migrants from northern mill towns, in the aftermath of the collapse of the British textile industry in the 1960s and 1970s.[10] Some of those who were made redundant in the northern mill towns moved south in search of work. As was common at the time, they were often employed as unskilled manual workers on low wages.[11] Even so, workers were able to sustain their immediate families and continue to provide remittances to relatives 'back home'. Pakistani migrants were the majority here,[12] mainly from the rural hinterland of Mirpur province.[13] Migrants also included a substantial number from the Sylhet region of present-day Bangladesh.[14] Both Mirpur and Sylhet were economies based on pastoral and agrarian pursuits, and both places remain overwhelmingly Muslim.

Malik was a sixty-four-year-old man who owned a grocery store on Dunstable Road. He first moved to Luton in 1973 not from Pakistan, where he was born, but from Bradford. He came to Britain in the mid-1960s and started work in a textile factory, along with other migrants who came from his home district of Kotli, in Mirpur province. After the factory was shut down, some of his friends moved south and found employment at the Vauxhall plant. Malik soon joined them.

> I worked at the factory until 1987, when I decided to open my own business. I remember so many Pakistanis were working in the factory in those days. It was a good living. We all earned enough to buy our own houses and bring over the families. [We also bought] land in Pakistan and supported our family members there.

Malik was particularly proud of the fact that he has supported (and continues to support) his family members both in England and Pakistan, and was keen to point out that such attitudes were changing among the younger generation. Younger Muslims are no longer willing to continue the burden of remitting their earnings to relatives they barely knew – a common subject of conversation and passionate debate among the youth with whom I spent time: 'we hardly know them, why should I give my money to them?'; 'they have it better than us'; and 'they'll never have enough' were some popular retorts.

In 2002, Vauxhall Motors closed its factory in Luton. Workers eventually lost their jobs, although the laying off of workers was staggered over a period of time in order to minimise impact. Many of the town's Muslims again faced an uncertain future. They were largely unskilled, with significant financial commitments, and with large families both in the UK and South Asia. In the absence of replacement jobs, a significant number of Muslim men (the primary breadwinners) sought self-employment, or diversified into the service sector. Taxi driving and small businesses ownership became popular options. Liaqat was a fifty-seven-year-old man and a taxi driver by profession. He used to work at Vauxhall Motors, but they laid him off in the 1990s. Since then, he'd been driving his taxi. He was happy with his job, as it paid well and suited his lifestyle. But Liaqat was morbidly overweight, suffered from Type 2 diabetes, high cholesterol and high blood pressure, all in addition to the indignity of supporting Manchester United. He claimed he had no time to exercise, play with his grandchildren or even watch his football team embarrass themselves every week.

> The trouble is, now that I'm driving the taxi, I'm having to work more than I ever did in my entire life! With cabbing, you are your own boss. You set your own hours. You don't answer to anyone. And you can make as much or as little money as you want. But the human being is greedy, no? So I now work on average a hundred hours a week, seven days a week. My wife tells me to calm down but I tell her 'who's going to pay for all the bills, and the weddings, and the mortgages?' My sons are still studying, so I'll stop when they decide to become men!

Liaqat worked hard for his family and, like Malik, regularly remitted back to relatives in his ancestral village in Pakistan. Over the years, his remittances have accrued vast tracts of agricultural lands, and paid for the construction of some lavish mansions. In Luton, he lived in an overcrowded three-bedroom terraced house. On average, he travelled to Pakistan once every few years and, being self-employed, could afford to take weeks and even months off at a time. Once in Pakistan, he managed his estate and relaxed in the mansion (*koti*) that he built for his brother, who remained in Pakistan. Whenever he tried to persuade his adult children to join him, they all declined. 'They don't like Pakistan', he said with a grin. 'They think it's better here. They don't trust them ... but I can't abandon my own'. Liaqat, like so many men of his generation in Luton, struggled to convince his children that keeping permanent links with South Asia is a good thing. Young Muslims with whom I spent time cited a number of reasons for their apathy towards their ancestral home.[15] Some claimed that their parents were being swindled by greedy relatives. Others were not impressed by the oppressive heat, mosquitos, power cuts, food and lack of running water. Some (especially women) also had issues with the culture, claiming they felt restricted and could not relate to their cousins' tastes and interests. Young people's attitudes towards Luton, however, could not have been more different.

Omar was a seventeen-year-old male of Pakistani heritage. He described himself as 'pure English'. This was unusual as he was the only respondent to self-identify as English. The vast majority were not comfortable with the term, as they claimed it denoted a white ethnic identity. Almost all of my informants, however, were comfortable with identifying as 'British'. When I asked Omar to explain why, he simply replied 'chicken and chips'. Bemused, I asked him what he meant by that, and after a little thought he replied:

> My mum and dad speak English to me and all my brothers and sisters. I grew up in Luton, it's my home. We go back to Pakistan now and again, but I hate it there. All my brothers and sisters complain when we're there. We complain so much that my dad got the message, and doesn't force us any more. Actually, he doesn't go as much as he used to. I'm not used to being over there. The people are different, they're all weird; they talk about things that I don't care about, and are always after your money. In Luton I have my friends, and my football team. I can understand the people here. What do I have over there apart from a huge house that nobody lives in, and goats and chickens? They don't even have a park where you can have a kick-around in. Nah, it's not for me. I prefer Luton.

Omar loved Luton. And he was not the only one. Almost all of the British-born Muslims whom I spent time with claimed that they were loyal to Luton before anywhere else in the world. Some of the more religious respondents argued that Luton was *the* place in the country to be a practising Muslim. They suggested that the high concentration of Muslims in the town, the number of mosques and seminaries (*madrasas*), and a state attitude of tolerance towards religious minorities meant that Muslims in the town were free to practise their religion without fear of sanction. This also meant that Muslims were able to live much better lives than some of their co-religionists around the world. Although this may be a somewhat romantic view given the subsequent emergence of the EDL in the town, it is one that was reiterated over and over again by 'practising' and 'non-practising' young Muslims alike. Kamal was a twenty-seven-year-old Salafi Muslim male whom I met at a mosque run by the sect. He was very observant and organised his life around the daily rituals. He could often be seen at the mosque, even when there were no congregational prayers or events happening, reading the Qur'an in the corner of the prayer hall, or in conversation with another devotee. Kamal claimed that Luton is second only to Saudi Arabia as a Mecca for the pious:

> Luton is full of Muslims. There are more practising Muslims in Luton than [there have been] ever before. You only have to walk around the streets to notice so many women wearing the *hijab* [Muslim headscarf] and men with long beards and dressing in Islamic clothes. There are halal shops everywhere; even in the schools and universities you can find halal food; this shows that even non-Muslims have accepted our way of life.
>
> [...] I've been to London and other big cities up north, and I never get the feeling of safety and homeliness as I do in Luton. Maybe because I was born and raised here, I don't know. But certainly, I find it much easier to be Muslim here than other places in the country. I don't even get this feeling when I go to Pakistan; people there seem to be far from Islam compared to here. The only other place in the world I would move to and feel totally comfortable in is Saudi Arabia.

Luton, then, is clearly not so horrible for its Muslim inhabitants. The 'elders' seemed satisfied with what the town had provided for them. Here, they were able to find employment, settle with their families, and get on in life. Their children and grandchildren, observant Muslims or not, also spoke well of the town and regarded it as their home. Some, like Omar and Kamal, never wanted to leave, albeit for varying reasons. Omar was happy playing football with his friends, going out with his White girlfriend, and the abundant supply of 'chicken and chips'. Kamal had all he needed in Bury Park: religious freedom in the form of mosques, halal food, and community spirit.

I ended up in Luton by pure chance. Originally (and rather naively), the idea was to have two field sites: one in Pakistan, and the other in Britain. The initial project was focused on the dissemination and flows of religious knowledge between the two countries, in order to assess to what extent religious devotion and attitudes shifted or remained the same in the two con-texts. Needless to say, the highly sensitive nature of the proposed research (involving working in many *madrasas* which, at the time, were accused of harbouring pro-Taliban terrorists and sympathisers), meant that securing a research visa was next to impossible. Given this set-back, the focus of the research became exclusively on Britain. The difficulty then lay in trying to find a suitable place to situate the study. In the first instance, I had planned to conduct the work in one of the northern mill towns that had housed so many Muslim migrants in the 1950s and 1960s. However, access proved to be too difficult to secure. An opportune moment arose, however, when an organisation based in Luton advertised for a position as a researcher on an Action-Research counter-radicalisation project funded by the European Union (EU). I figured that the position would provide an ideal gateway to gain access to young Muslims, who were the target group for the project. I applied and, fatefully, was offered the position.

The project was run by a group of second-generation Salafi Muslim men in their thirties and forties of South Asian heritage who were (and remain) particularly active in Luton. It was a sister organisation to the central Salafi mission in the town. The latter was based at an old syna-gogue in Bury Park that had been converted into a mosque in the 1990s, and where I spent a lot of time. There I either shadowed my key inform-ants from the EU project conducting interventions and holding public meetings, or met worshippers in my own time and capacity. The coun-ter-radicalisation project – which I will call the Minority Skills Project (MSP) – worked with a number of youth centres, schools, the local police force, Luton council and the Salafi mosque. Its aim was to iden-tify vulnerable young men and discourage them from joining extremist organisations through theological interventions. The project had been established in 2000, and worked primarily with Black and Asian men who were ex-offenders returning from prison as well as with convicted criminals serving time within prisons, drug and alcohol addicts, and unemployed youth, helping them to re-integrate into society. Officially (and in the eyes of its state funders), the project involved the provision of advocacy services and vocational training, assistance with drafting CVs and job applications as well as providing counselling. In reality, it was a front for proselytising young Muslims (and some non-Muslims) into the Salafi sect. The organisers used interventions in prisons, schools, youth centres and at their offices to convince young men that a life of piety ulti-mately yielded more rewards than a life of petty criminality, imprison-ment, substance abuse and indolence.

I first met the project leader, Fakruddin, when he picked me up in his car from Luton train station. It was the day of my interview. He was a kind man, and rather eccentric. When I first laid eyes on him, I was quite surprised. Prior to the interview, I had no idea that the MSP was run by Salafis. When a small, dark man, got out of his car, wearing a very large unkempt beard, skull cap (*topi*) and white *salwar kameez*,[16] I was quite taken aback. I instantly recognised the garb and grooming as hallmarks of those belonging to the sect. I was quite perturbed as I thought this may cause problems if I got the job. In my experience, Salafis were intolerant of all Muslims that did not belong to their sect, and viewed the act of inviting (*dawa*) others to the sect as a fundamental religious obligation.[17] Being a Muslim myself, albeit one that did not observe all the daily rituals, and belonging to a mainstream school of Sunnism, I was concerned that they might try to convert me. Fakruddin greeted me warmly with a shake of the hand and a robust hug. While we drove to the office, he formally introduced himself and briefly outlined the aims and objectives of the project, punctuated by the occasional pointing out of the sights of interest in Luton.

Fakruddin was an interesting character, and quite unlike any other that I met during my time in the field. He graduated from the London School of Economics in the late 1980s, where he attained a degree in Accounting and Finance. During his time at university, he was an active member of the Socialist Workers' Party, and even spent a stint in jail for allegedly being involved in a violent confrontation during a demonstration against the Poll Tax in 1990. In contrast to the other staff of MSP, however, he was not from Luton. Fakruddin lived in St Albans, a town in the neighbouring county of Hertfordshire. After finishing his degree, he moved back in with his parents. He claims it was after he moved back that his transformation from radical socialist activist to a politically quietist, puritanical Muslim began. He had been charged for his involvement in the Poll Tax riots and was awaiting trial. If found guilty, his chances of securing a respectable job would be severely compromised. During this period, he was deeply concerned for his parents' welfare, as they had worked hard to provide him with a decent education. According to Fakruddin, the strain of the looming trial affected his parents' health. While awaiting the trial, he started attending the local mosque and following regular visits he eventually became acquainted with a member of the congregation. Fakruddin knew this individual from childhood but had previously avoided him as he was always preaching to the youth and applying pressure for them to live a more pious life. An older, weathered, and politically wounded Fakruddin was more open to his call. After a while, he accepted an invitation to attend a lecture in a mosque in Luton. There, he met other members of the British Salafi community, including his current colleagues and many of its prominent leaders and scholars. He unreservedly cited these meetings as the auspicious turning point of his life, as it was then that he not only 'discovered Islam' but, due to his earnest repentance and pious acts, God relieved him and his family of their turmoil. Fakruddin's case was dropped in court for lack of evidence and he was duly acquitted of any wrongdoing. By the time I met him, he was the most dedicated member of staff at MSP, which he also co-founded. His transition from an activist community of 'workers' to 'believers' was thus complete.

Once we arrived at the offices of MSP, I was formally introduced to my team members for the research project. They included Ammar and Khidr. Ammar was in his mid-thirties, and Khidr was in his early forties. Like Fakruddin, they were both married with children, and both of them (as well as all the other staff in the office) were male. They all had very long beards, and dressed in either traditional Pakistani clothes or in military/outdoor clothing and wore trousers that were too short for them as a display of modesty (all typical of British Salafis). Like many others working in the project, Ammar was a reformed local gangster who was now an experienced youth worker. Over the years, Ammar had also accumulated an impressive resume of vocational qualifications related to health and social care. He was a charming man: gregarious, generous and fun-loving. Over time, I realised he was enormously popular among all ages in Luton and its immediate environs. Wherever we ventured in and around town, we were bound to meet someone who knew him. And it made sense. Not only was his open character inviting, but he had reached the pinnacle of both criminality and piety in his lifetime – two areas of activity that commanded the most respect among the young Muslim men with whom I worked.

Khidr was the veteran of the gang. He had been a Salafi missionary (*da'ii*) since the early 1990s, and had spent years in Saudi Arabia training as a jurist under eminent Salafi scholars. He was fluent in classical Arabic (*fusaha*). Khidr was also very likeable. He was hospitable, amiable and highly charismatic – especially in large gatherings. Unlike the others, Khidr was married to a White convert to Islam, having been previously married to a Black convert of Caribbean origin. He had numerous children from both marriages, and was also significantly wealthier than his colleagues. He lived in a more affluent part of town, drove a luxurious, speed-hump-crushing 4x4, and owned a number of businesses. Whereas most of Luton's Pakistani diaspora originate from Pakistani-administered Kashmir, Khidr's family were originally from northern Punjab. Interestingly, although he was highly proficient in speaking, reading and writing Arabic, his Punjabi lingual skills were non-existent and he spoke with his Punjabi-speaking parents in English. When probed as to why this was the case, he replied that he did not see the merit of speaking Punjabi when he was growing up as he considered himself to be British, not Pakistani. It was only when he became more interested in Islam that he decided to learn Arabic, a language that he claimed was integral to understanding Islamic scriptures.

Khidr was the senior *khatib* (public speaker) at the Salafi mosque. This meant that he often delivered the Friday sermon (*khutba*). This was the mosque where Fakruddin had experienced his epiphany, and it was the epicentre of social, educational and devotional activities for Luton's Salafi community. A private Muslim primary school was also located on the premises, where Salafi parents from in and around Luton sent their children. In addition to national and international conferences that were held at the mosque (and where keynote Salafi speakers were invited from around the world), the space was also used for weddings and other social events. It was not uncommon to find scores of men socialising in clusters in the prayer hall, or loitering around the main entrance after congregational prayers. Among the congregation, Khidr was a highly respected and influential individual. He was famed for his knowledge of religious sciences, oratory skills and debating prowess. He was also lauded for the personal sacrifices he has made for the mission (*daw'a*). His influence also extended beyond the mosque and the Salafi community. He had frequently interceded in and arbitrated disputes between Muslim gangs engaged in turf wars for control of the lucrative and burgeoning drugs trade in the town. His piety and status as a pioneering and financially successful local Muslim wielded considerable authority among the younger generations, many of whom felt compelled to honour his wisdom. Khidr was the intellectual heavyweight among the town's Salafi contingent, and many of the MSP's ideas and initiatives were directly inspired and implemented by him.

The interview commenced with formal introductions and a brief outline of the proposed project. Ammar would act as the Youth Liaison Officer, which entailed contacting teachers, youth workers and imams of mosques and informing them of the work and arranging interventions. Khidr would be the key speaker and interventions coordinator. He was responsible for creative planning and delivery of intervention sessions. Fakruddin would manage the entire project, including budgeting and logistical support. My role was to attend all the interventions, take minutes and make detailed notes of each session. I was also responsible for writing the report to the EU at the conclusion of the project. For the next year, I spent a significant amount of time with these men. My initial concerns were allayed during the interview after it became apparent to me that they were pretty desperate to recruit a Muslim male with formal qualifications in the social sciences. They did not seem too bothered by my religious views, and I felt assured that they would not try to convert me, which was a huge relief. Fakruddin, Ammar and Khidr became my chief informants in Luton and, through them, I accessed a large portion of my subsequent respondents. The widespread respect that they were afforded in the community meant that being associated with them immediately endorsed the credibility of my own work, and that made my task a lot less laborious.

***

Over the next year, I spent a lot of time with the team at MSP, especially with Khidr and Ammar. On some occasions, we would venture beyond Luton to other parts of the country, on camping residentials or meetings with project stakeholders, for example. Mostly, however, we spent our time in Luton. Our work was based at six different intervention sites: one high school, three youth centres, the Salafi mosque and the offices of MSP. Sometimes I attended interventions at local restaurants too. At the schools and youth centres, Khidr addressed large groups of mixed Muslim youth, aged between fourteen and eighteen. At the youth centres, we usually addressed boys, due to the relatively smaller number of Muslim girls attending our respective clubs. The age range at youth centres was also broader – we engaged with children from as young as eleven up to young adults in their late teens. At the MSP offices, interventions with older men were conducted. Sometimes, we even took men in their thirties out for dinner.

One such man, Lala, was a crack cocaine addict. He was in his late thirties, extremely slender and had been divorced twice. He was a sensitive, personable man, who really liked to chat. At the time of research, Lala was dating a White woman who was also the mother of his fifth child, and her fourth. He was married to his cousin in Pakistan when he was in his teens. The marriage was arranged by his father's older brother and patriarch of the extended family (biradari). Lala claims he tried his best to make the marriage work but, in his own words, they were 'just different'. He was outlawed by his uncle following the divorce, and forbidden from travelling to his ancestral village in Pakistan. But Lala didn't seem too bothered: 'that was the best part!', he quipped. His second marriage was to a fellow British-Pakistani; a woman from Manchester with whom he shared a home for several years. But that marriage fell apart because of crack, he claims. Lala was a builder, painter decorator, and general handyman. On occasions, when he went into remission, he would find casual work at building sites for cash-in-hand. During these periods, he claimed to provide full maintenance for his children. Most frequently, however, he went missing for days. 'It's ruined my life, but I'm weak', he told me. Lala felt very guilty for being an addict. He thought it was a grave sin (*gunaa*) to indulge in drugs and alcohol but couldn't help himself. He said it made him feel better, by not feeling at all. The team at MSP came across Lala at the Salafi mission. When he felt depressed, he would often stroll into the mosque and sit in solitude in the prayer hall. If there was a study circle (*halaqa*) in progress, he would sometimes join the seminar as an observer. Khidr often led such sessions and – never missing an opportunity to preach (*da'wa*) – struck up a relationship with him.

I first met Lala on a weekend residential trip to an outdoor activities centre in Staffordshire. He sat next to me on the minibus from Bury Park and talked the entire way. He told me he decided to come on the trip at the very last minute after his girlfriend had pressured him. She was not Muslim but she thought religion would help Lala give up drugs. She was at the end of her tether and he was keen to make amends. Although Lala was not at risk of radicalisation, he was being groomed for recruitment into the Salafi sect. Khidr had 'converted' many such individuals in the past, and the congregation included a disproportionate number of reformed ex-addicts. Whatever formula or antidote Khidr offered, it certainly worked for some. Lala, however, slipped in and out of piety, depending on his mental and emotional state.[18] Although this was to be a dry weekend, Lala had other ideas. His suitcase contained a large bottle of Irish cream, a bottle of vodka and a huge bag of marijuana. He also brought a few crack rocks. I could see why he insisted on sharing a room with me. All the others on the trip were either teenagers or teeto-talling 'practising Muslims' who would not only disapprove, but be highly offended at the mere sight of drugs and alcohol. Lala didn't sleep. At one point, around three in the morning, I woke up and the light was on. There was also a pungent smell in the air. I looked down into the bottom bunk, and Lala was smoking a pipe. When he noticed me, unperturbed, he caus-ally stretched out his arm and offered me the pipe. 'Didn't think you'd want some of this', he said after I politely declined. And so he carried on.

Lala was the cause of much shame (*sharam*) and dishonour (*besti*) for his family, especially his parents. Divorce and substance abuse is a great taboo among the older generation, and Lala's condition was an open secret for those who knew him. His parents struggled both with with having an addict for a son and having everybody know about it. Lala was also aware of his parents' suffering and it affected him. Over the years, he had tried different kinds of therapy and counselling, but nothing seemed to have worked. Being with 'the brothers',[19] he claimed, gave him the most comfort. Lala stopped coming to the interventions after a while. The last I heard, he had left his girlfriend and moved to Manchester, where his ex-wife lived.

***

At this juncture, it is probably necessary to declare that, like some of my informants in Luton, I am also of Bangladeshi origin. This admission obviously complicates the dynamics of this work. Studying a group of individuals with so many similarities with my own formative development, therefore, is a major methodological and ethical tension of this work.[20] Like many of the families now settled in Luton, my family also arrived in the UK as a direct consequence of the 1971 Immigration Act. My father came to Britain in the mid-1960s as an economic migrant. Once in Britain, he lived for a time with his older brother in Sheffield, South Yorkshire. My uncle had arrived in the late 1950s, and took up employment as a crane operator for British Steel. Like many others of their generation, my father and uncle never intended to stay. Again consistent with their generation, both men had left their wives and children in Bangladesh (East Pakistan at the time), hoping to re-join them eventually. In 1982, after years of lobbying, my mother finally managed to convince my father that moving to England was a good idea. Top of her list of reasons was the far superior opportunities available for children in Britain compared to newly independent Bangladesh, which was at the time a country ravaged by poverty and famine, natural disasters and endemic political corruption.[21] Once in Britain, we eventually settled just beyond south-west London, in the town of Kingston upon Thames, Surrey. This was an unusual place for a Bangladeshi family to move to as very few Bangladeshis (or other ethnic minorities) lived in that part of the country. Consequently, my early life was spent in a town with a majority-White population. In contrast, and totally unfamiliar to me personally, my informants in Luton had been raised in an area where the majority of the local community shared a common ethnic, religious and historical background.

At the time of migration, I was eleven months old. I had therefore no memory of my country of birth, unlike my three older siblings. Like my friends in Luton, Britain was the only home that I knew. Despite this, throughout my childhood and adolescence, I knew that I was 'different', and that I didn't quite 'fit in'. Moreover, my family and I were frequently subject to racist abuse – both verbal and physical. The council estate in which we settled in those early years did not seem hospitable to South Asians. Our home was regularly vandalised by our neighbours: ranging from graffiti being sprayed on our walls ('Pakis Go Home', 'Black Bastards', 'NF' etc.), to human faeces and fireworks being dropped through our letterbox. On one occasion, in the late 1980s, I was being taken to school in the morning by my mother. A young man approached her, spat at her, and told us that we were not welcome there. At school, I was constantly reminded by my peers that I was 'Black' and a 'Paki' and that I should 'go back to where I came from'. In my teenage years, I was victim to a number of unprovoked violent street attacks by neo-Nazi 'skinheads' in my local area. During one incident, I was dragged across the street on my back, the scars of which I still carry. Thus, during my formative years, Britain seemed like an inhospitable place. Although I always thought of myself as British, I was always aware that to some this was unacceptable. Moreover, such experiences only enhanced my perception that I was 'Other' in society and that my skin colour and (to a lesser extent at the time) my religion were barriers to full integration. Being *really* British to me meant being White and Christian.

I learned that the young Muslims in Luton were not so different to me, twenty years on. Like me, they considered the English town in which they grew up to be their home, despite their darker shade of skin and 'exotic' religion. They were also all too aware of their minority status in Britain. Moreover, they were aware that there was a tide of antagonism by some sections of society towards them, due to their Islamic cultural roots. Despite this, for many, self-identifying as 'British' was curiously natural.

For all the similarities, however, there were some significant differences between myself and my young informants. The majority were of Pakistani heritage. Along with English, they often spoke Punjabi, Patwari, or Pahari with members of their kin, particularly with elders. I was not always familiar with particular customs and attitudes which had their origin in Pakistan. My first language is Sylheti, a regional variant of Bengali. Bangladeshi cultural norms and attitudes are, in many areas, significantly removed from those practised in Pakistan. Additionally, my informants found it difficult to understand why I had chosen to study them. I was often advised by both older and younger members of the community to get a job that paid well, or to think about becoming a doctor or a lawyer once I had finished my degree. On one occasion, a young man whom I had met at a youth centre asked me why I was 'doing a job that White people from the council did?'. Even though I shared the same religion, skin colour and minority status as my informants, they regarded me more often than not as an outsider in Luton. In this sense, my own pre-fieldwork perception of being a 'native anthropologist' was somewhat challenged. It occurred to me that I might share some superficial demographic similarities with the community I was studying but, ultimately, I was as much of an outsider here as I was in my own home town in Surrey, albeit for different reasons. Unlike my informants, I had never lived in a close-knit South Asian community where the majority of my peers came from the same ethnic or religious background. Moreover, my family had always encouraged education for its own sake over purely economic pursuits. In Luton, the situation was a little different. The vast majority of my informants saw education as a means of yielding more income in the job market, or a waste of time altogether. They were consistently bemused as to why I had spent so many years studying only to eventually earn less than the average taxi driver in the town.

The experience of fieldwork in Luton was thus an illuminating experience for me. I discovered that I shared many similarities with the people I studied, but also some overwhelming differences. Despite initially revelling in the prospect of 'studying my own', I was soon questioning my hitherto taken-for-granted status as an 'insider'. My informants were always hospitable, warm and welcoming. I always felt assured that I was regarded as just another member of the community. However, deep inside, as my time in the field continued, it dawned on me more and more, that I wasn't. I was not a participant in the process of community but merely an observer. And, no matter how much I wanted to believe it, my informants, generous and forgiving as they were, knew it too.

Notes

1. See Castles 1987; Solomos 1993; Ballard 1994; and Ansari 2004.
2. According to the 2011 National Census, Luton's overall population is approximately 203,200. The largest ethnic group in the town is white, with 44.6% of the population. The Asian British comprise the largest ethnic minority with 26.3% of the population, with the black British forming the second largest ethnic minority with 9.8% of the town's overall population. Persons of Pakistani origin number 14.4% of the population – by far the largest minority group in the town. Persons of Bangladeshi and Black African origin form the town's next largest ethnic communities; comprising 6.7% and 4.5% of the population respectively. In terms of religion, 49,991 or 24.6% of the overall population of the town were Muslim; of which the most significant were 29,353 of Pakistani origin and 13,606 of Bangladeshi origin. With regards to employment, 53,271 of the town's population between the ages of 16 and 74 were working full

time; 18,407 were in part-time employment and 12,198 were self-employed. There were also 8,059 full-time students and 8,337 unemployed. Of those who were economically inactive, 10,703 were students; 5,496 were permanently sick or disabled; 9,660 were looking after the home or members of the family; 14,336 were retired; and 4,741 were inactive in other ways. Of those who were unemployed, 2,097 were aged between 16 and 24 years; 1,340 were aged 50 and over; 1,648 have never worked; and 3,186 were long-term unemployed. The majority of those in employment were in the wholesale/retail trade sector, that is 17.8% of the overall population; 10.8% were employed in the human health and social work activities sector; 12.6% were employed in real estate, renting and business activities; and 10.1% were employed in the transport, storage and communication sectors. Within the Asian population, 21.9% were employed in the hotels and restaurants sector; 14.3% were employed in manufacturing; 14.1% were employed in the wholesale and retail trade and the repair of motor vehicles; 13.9% were employed in transport, storage and communications; and 12.1% were employed in mining and quarrying. With regards to education, an investigation is currently pending on the accuracy of student information with the Office for National Statistics and thus these figures could not be included within this study.

3. See Treadwell and Garland 2011; Barlett and Littler 2011; and Pilkington 2016.
4. Salafism, also referred to as Wahabism, is a puritanical, literalist sect of Sunni Islam that originates from eighteenth-century Arabia. It is the official religion of the Kingdom of Saudi Arabia, as well as neighbouring Persian Gulf countries, see Wiktorowicz 2006.
5. See Taji-Farouki 1996; Mandaville 2001; Roy 2004; and Wiktorowicz 2005.
6. An ethno-linguistic group originally from southern India.
7. Compare with Bhachu 1985; Shaw 1988; Eade 1989; Werbner 1990; and Baumann 1996.
8. See Castles 1987.
9. See Holden 2003.
10. See Kundnani 2001.
11. See Allen 1971.
12. Prior to independence from Pakistan in 1971, present day Bangladesh was known as East Pakistan. In the period in question, therefore, Bangladeshis identified as Pakistanis.
13. See Saifullah Khan 1977; and Anwar 1979.
14. See Gardner and Shukur 1994; and Gardner 1995.
15. Compare with Cressey 2006.
16. The national dress of Pakistan. It is a two-piece outfit consisting of very baggy trousers, and a long, loose shirt draping down to the knees.
17. See Mahmood 2005; Robinson 2008; Meijer 2009; Schielke 2009; Hamid 2009; and Inge 2017.
18. Compare with Schielke 2009; Simon 2009; and Marsden and Retsikas 2013.
19. This is an affectionate term that refers to other Muslim men, and is commonly used in Luton to refer to co-religionists. Muslim women are often referred to as 'sisters'.
20. See Oakley and Callaway 1992.
21. See Van Schendel 2009.

2
Family

Luton's Muslim community mostly originate from South Asia. Every young man I knew was a descendant of a pioneer economic migrant who made the journey to Britain in the post-war period. Although some of the pioneer migrants have since returned home (once their financial needs were met), most remained and permanently settled. The decision to remain, however, sparked a curious acculturation process, involving intergenerational negotiation and, sometimes, conflict.[1] Although first-generation migrants had opted to stay, they also decided to maintain acute links with South Asia. Some of these links were financial, mostly involving migrants sending back remittances to relatives for upkeep, or for investments in land and property.[2] Others were physical, with migrants periodically returning with their families for ancestral tours.[3] In the other direction, however, cultural norms were also exchanged. Community 'elders' were keen to enact South Asian etiquette, social relations and moral/religious instruction within their households in a British setting.[4] Whereas these norms were broadly accepted and expanded upon by many within the second generation – a significant proportion of whom moved to Britain as children or young adults – subsequent generations that have been born and raised in Britain are challenging the legitimacy of their heritage in providing a suitable cultural framework through which to live ostensibly 'British' lives. This chapter will aim to draw out some of these generational tensions and show how young Muslims successfully manage expectations from home alongside pressures from wider society. The chapter will demonstrate the way in which young Muslims accept, reject or hybridise discursive influences from both the 'South Asian home' and 'White liberal society' to create unique identities and subjectivities that succeed in redefining established notions of what it means to be British in the twenty-first century. The chapter will chart the generational evolution of the community, bringing in voices from three generations of the town's Muslim community. It will provide ethnographic insights into how early migrants settled in the town, their encounters with racism and exclusion, and the ways in which community solidarity helped them overcome those early challenges. These accounts will be contrasted with how British-born Muslims manage to juggle (or not) expectations from home and those from the outside world.

***

I first met Omar during an intervention at a mobile youth centre in Bury Park. This was one of the three youth centres that Khidr, Ammar and I regularly attended in search of potential 'terrorists', in order to 'disarm' them with theological lectures. The mobile youth centre itself was a converted lorry. Inside, there was ample room to move around quite freely. At the rear of the cabin, there was a seating area, complete with cushioned seats and a modest table. The rest of the cabin contained a sequence of flat-screen televisions attached to the walls and a corresponding chair situated in front of each screen. Each chair was occupied by a young Asian male, holding controllers and acting in a generally boisterous manner. They were playing a video game on the latest PlayStation called 'Call of Duty' – a violent military first-person 'shoot' 'em-up' which one can play against the computer, as a team or against each other. These young men were aged between sixteen and eighteen, and they all lived in the estates that surrounded the recreational park where the lorry was anchored. In addition to providing space for seasonal sports, such as cricket and football, the park was notorious for being a centre for anti-social behaviour. Scores of youth could be seen loitering there at any time of the day. Some of them indulged in drug-taking, some dealt them, while others just came and played football. The majority of them were young South Asian males. Occasionally, however, some young women did dare to venture into the park but, if they were not known to the young men, they were usually heckled. Some of this heckling was very aggressive and sexual in nature. The young men in the youth centre, as I later discovered, were a few of the chief culprits.

The youth workers ordered the young men to conclude their game and, once they did, they rather reluctantly joined us at the back of the lorry. The topic of the intervention was 'Is Britain a Land of Peace or a Land of War?'. Quite bemused, the young men asked Khidr to clarify the question. Khidr explained to them that these terms were taken from the Islamic Scriptures, and that they were terms to define Muslims' relationships towards a foreign land, i.e. countries that were not majority-Muslim. Omar immediately replied 'How can it be a land of war?'. Khidr explained that some people think that Britain is a legitimate target for attacks, like the 7/7 bombings, and asked the group what they thought about that. Omar responded again: 'How can it be a land of war; we live here? It's dumb'. At the time, Omar was seventeen years old and very outspoken. He lived with his family a few streets away and spent most of his time in the park with his friends when not at college. Omar had one older brother and one younger sister. His parents were both born in Pakistan but settled in Britain when they were children. As we know from the previous chapter, Omar considered himself to be 'pure English' and had an insatiable desire for chicken and chips – Bury Park's most famous culinary contribution (Dunstable Road and the surrounding area is crammed with competing halal fried chicken outlets, each with its own unique spicy sauce). Omar also had a girlfriend of whom he was very fond. She was White, and they often met in the park. Omar's parents didn't know about her and he was keen to keep her a secret. When I asked him why, he said because his parents wouldn't approve of him having a girlfriend. His sisters, however, did know about her, but kept it a secret from the 'elders'. 'That's why we meet here, or outside of Bury Park', he said. 'Don't want anyone to see us. No one comes here apart from people our age who are doing the same thing'. Although, in Omar's words, his parents aren't 'typical Pakis' and would probably accept a White girl if she converted to Islam, they were still strict in 'certain things', he claimed. Because of this, Omar had to delicately manage his social life alongside his home life, often suppressing his opinions and rebellious character:

> There's a massive switch when I get home, I've got to pretend to be someone else. If I told my mum that I had a girl, she'd disown me! I got to hide a lot of stuff [that] I get up to – normal stuff for *Gorreh* [Whites], like having girlfriends, drinking, going to clubs. But Pakistanis can't do that stuff because it's *haram* [forbidden in Islam] [...] I think that's all bollocks [because] every Pakistani does it, they just hide and do it [because] they don't want people to know about them [because] it will look bad on them and their family.

Omar's speech was laced with Punjabi words, most of his friends were Asian, and he claimed to never miss Friday prayers (*Juma'a*) at the mosque. Despite this, he could not speak Punjabi with any level of fluency, could barely read the Qur'an, and spent most of his free time smoking cannabis in the park – sometimes in the company of his long-term non-Muslim White girlfriend. However, when I asked whether he'd be happy for his sister to have a boyfriend, he got into a bit of a frenzy:

I'd kill her! Then I'd kill him! I don't want her to have a boyfriend. I know what boys are like, they will take advantage of her. It's wrong for Pakistani girls to have boyfriends. It's not right. Especially not my sister.

[...] Alright, maybe I'm not 100% English, but that's the only thing! [*laughs*]

***

In 1971, Britain changed its immigration laws. Prior to this, unskilled workers from the ex-colonies were permitted free access to the UK. However, fierce debates raged in parliament in the 1950s and 1960s that eventually led to the Immigration Acts of 1962, 1968 and, finally, 1971.[5] Although the 1971 amendment meant that New Commonwealth citizens were no longer permitted unrestricted entry, it did allow migrants that were already resident in Britain the right to bring over their immediate families. This, obviously, directly affected the thousands of New Commonwealth migrants – mostly men – who had moved to Britain to answer the government's call for labour in the aftermath of the Second World War. They had witnessed a radical shift in their residency status over the course of nine years. Ever pragmatic, South Asian migrants quickly moved to encourage kinsmen and fellow villagers to migrate to Britain before the restrictions came into place. Thus, instead of restricting immigration from the New Commonwealth, the 1960s witnessed an explosion of 'chain-migration' prior to the Act coming into place.[6] Initially, migrants already in Britain sponsored kin and fellow villagers to make the journey for work. They remitted funds for the procurement of a passport and ticket for the voyage. Once in Britain, new migrants were provided with accommodation and work by their sponsors. This not only created areas in South Asia where there was a high concentration of 'out-labour' to the UK,[7] but also facilitated the emergence of distinctly South Asian areas within British inner cities and towns in close proximity to various industries.[8] Secondly, with the introduction of the 1971 legislation, the South Asian population of these areas were further bolstered with the arrival of wives and children throughout the 1970s. This seminal period led to the eventual consolidation of the presence of South Asian and Caribbean communities all across Britain, not least in Luton.

For those early migrants from South Asia, not only was the 'myth of return' permanently dispelled by the arrival of their families,[9] but it also meant that the children of first-generation migrants were to be schooled and socialised in a cultural environment that was completely different to that of their parents and ancestors. Fearing a loss of their 'culture', first-generation migrants sought to re-apply South Asian customs in diaspora, primarily within the home.[10] Interactions within the household were expected to be conducted with appropriate decorum depending on age, domestic rank and status, as was common in households in South Asia.[11] Within South Asian families, individuals were expected to be primarily loyal to the collective interests of the group – in practice, this usually translates into obeying 'elders' and performing deference. That is not to suggest that they had forsaken or repressed their own individuality or agency,[12] but that their sense of personhood is intimately intertwined with group solidarity and family honour.[13] Consequently, some scholars argued that the second generation were caught 'between two cultures' and were resigned to juggling the conflicts and contradictions that came with honouring their cultural heritage at the same time as being 'British'.[14] In addition to conforming to kinship hierarchies and codes of honour, other cultural conventions were typically played out in the family home through the observation of ceremonies and social events[15]. To a large extent, British-born Muslims in Luton today continue to be socialised within this household structure. While, some forty years later, this conception still retained some resonance among my informants, British-born Muslims have developed novel ways of self-identifying that valorise their ethnic and religious heritage[16] while, at the same time, creatively situating such identities firmly within nationally accepted multicultural discourses.[17] Young Muslims in Luton saw no contradiction between being originally South Asian *and* being British.[18] Curiously, on the issue of religion, every young Muslim claimed that Islam does not necessarily belong to any particular country, but can be observed and practised anywhere, including Britain. This 'de-territorialisation' of Islam is consistent with youth attitudes seemingly pervasive in diaspora settings in the West,[19] and is a discussion that I shall return to in the final chapter of the book. For our purposes here, however, it is important to note that although young Muslims self-identified as British, they had nevertheless been uniquely socialised to recognise and negotiate hybrid and ambivalent moral and cultural registers in their everyday lives.[20]

***

Pre-war migration to Britain from South Asia was usually confined to seamen (*lascars*) 'jumping ship' at British ports and seeking work as casual labourers in the booming catering and manufacturing industries. There was plenty of work to go around in those times, and migrants were generally welcomed by the 'host' population.[21]

> Coming to Glasgow, 1937, I run away from Arcade ship, to London. Other people telling, 'London very good'. That time, England very good, people very respect colour people.

> – *Mr Sona Miah, Bangladeshi migrant from Sylhet*

> When I came [to Liverpool in 1944], I had nowhere to go – no address, nothing. I was sitting in the train – very quiet and miserable. One lady and gentleman were there with their child, and the child said, 'Look Mummy, there is a black man!' Black man! You know, little kiddies – they say whatever comes into their minds. The mother kept saying, 'Stop it, he is a gentleman!' Then he would stop for a while, then again start. Then I said, 'I don't mind, he is only a little kiddie, and after all my skin is black.' She said, 'Oh – thank you very much.' The father gave me a cigarette, and asked me where I was going, and I said, 'Birmingham.' After get out in Birmingham, they ask me if I got address, I say, 'Yes.'

> And show address. They say, 'You wait here in front of the station, while I take my wife home, then I come back.' He came … he brought a tram fare … he took me to house where I got address – one of my village men. One English lady opened the door and said, 'You want Karamat? Come on in.' Then I sat down and that gentleman gave me ten shillings, said, 'Here you are – buy cigarettes.' Then he went.

> – *Mr Syed Rasul, Pakistani migrant from Mirpur* (Adams 1987:137; 184–5)

With the mass influx of migrant men in the 1950s and 1960s, the subsequent unification with their wives and children, and the demise of manufacturing industries, attitudes began to change. Pioneering migrants, such as Mr Miah and Mr Rasul, entered Britain as subjects of the British Empire. Although, technically, they did not have permission to work – as they had absconded from their contracts with respective shipping companies – immigration and employment laws at the time were much more relaxed. Over time, they were naturalised with relative ease. The early welcome, however, turned to hostility throughout the 1960s and 1970s as the hard impact of economic decline set in. In the beginning, migrant men lived and worked together. They lived in lodging houses, and took up jobs that the native population refused. Lodging houses were overcrowded and luxuries were out of the question. Malik, now a successful businessman, recounted those early years to me with manifest pride:

> Eighteen of us shared [a] house and slept in the beds in two shifts. We all worked in the same factories. The day shifters would sleep in the beds during the night, and the night shifters would sleep in the same beds during the day [...] When I first came, I was working night shifts, and it took me a long time to get used to the routine. Even on my day off, I was expected to wake up at a certain time, because the day shifter would want to sleep [...] there was no space anywhere as all the beds were full of bodies [...] My house was full of people from Kotli[22] [...] Those days were tough, but everybody lived like that. It was the best way to save money, and send it back home. If I didn't live like that in those early days, I wouldn't have been able to bring my entire family to England.

Pressed by time and charged by the knowledge that they would return home one day, migrants worked as many overtime hours as they were offered. Once they had enough money, their sojourn would be over. Of course, it did not quite transpire this way. Immigration restrictions, the loss of jobs, and demands from relatives to continue to remit, all pointed towards settlement. As a single man, it did not cost much to survive. As a family, however, the situation was altogether different. The unification with wives and children led to a demand for housing and, with many migrants unable to afford a home outright due to their low wages, they applied for social housing.[23]

Early migrant men mostly resided in urban areas, in close proximity to their place of employment. Due to their relatively low income, homes were rented in deprived neighbourhoods and wards already occupied by similarly impoverished 'natives', who also worked in the same industries. While the going was smooth, relations between workers were, on the whole, convivial. However, when wives and children began to arrive from the 1970s, these relations took on a more sinister complexion. The arrival of families coincided with a loss of jobs for natives as well as migrants. Thus the pressure on the welfare state in this period reached chronic proportions. Migrants and families were particularly affected by institutional racism on the part of the local authorities. The burgeoning Bangladeshi community in Tower Hamlets, for example, faced discrimination on both an individual and institutional level throughout the 1970s.

> There was an urgent need for more accommodation, and the Bengalis faced more than their share of problems. First there was discrimination in council house allocations. Much of this was built into the system, but a lot was due to conscious or unconscious prejudice and assumption. An independent report commissioned by the GLC [Greater London Council (1965–86)] in 1983–4 gives a picture of deep-seated racism in the housing bureaucracy; a view confirmed by the squatting activist, Terry Fitzpatrick. He recalls accompanying a Bengali to request a transfer away from the terrorism of racist neighbours, only to be told that the requested empty flat was 'for white people'. (Glynn 2004:10)

The council homes that were eventually offered to them were situated in outlying areas of the borough, predominantly occupied by white working-class communities.[24] In a period of popular racism towards immigrant communities, spurred on by xenophobic right-wing parliamentarians such as Enoch Powell and his infamous 'Rivers of Blood' speech, isolated Bangladeshi families faced hostilities in all-white housing estates. For most, living in such conditions became untenable:

> The danger of racist violence on outlying white estates meant that re-housing in the Spitalfields area came to be seen as a matter of survival. As a contemporary report pointed out, the housing officers' original biased allocations reinforced this racism because they 'gave those white families the feeling that they had the 'right' to keep their estates white.' Many Bengali families who were allocated council housing on white estates returned to Spitalfields preferring to face the extreme discomfort of a squat to the constant danger of racist attack. (Glynn 2004:10)

The picture was similar for South Asian communities in northern mill towns. In Bradford and Oldham, the phenomenon of 'white flight' meant that those natives who could afford it moved to outlying suburbs as soon as mass migrations to the city commenced. Like Tower Hamlets, those natives who sought social accommodation, were prioritised by the predominantly White council staff. In a relatively short period of time, these structural machinations led to the consolidation of segregated neighbourhoods, divided on the basis of ethnicity in dozens of cities and towns across Britain.

The fear of racial harassment meant that most Asians sought the safety of their own areas, in spite of the overcrowding, the damp and dingy houses, the claustrophobia of a community penned in. And with Whites in a rush to flee the ghettoes, property prices were kept low, giving further encouragement to Asians to seek to buy their own cheap homes in these areas. It was 'white flight' backed by the local state. The geography of the balkanised northern towns became a chessboard of mutually exclusive areas. (Kundnani 2001:107)

Bury park, it seems, evolved in a comparable manner. Through the combined processes of 'chain migration' and 'white flight', culturally distinct areas of Britain's post-industrial towns and cities took shape, and became the ethnic enclaves of today. With the burgeoning population, however, came the need to fortify economic pursuits. Social networks therefore – and, in particular, ties with kinsmen and benefactors – became all the more important in the face of an ailing labour market. Such networks were instrumental in generating capital for the opening of businesses, the buying of houses,[25] and the building of community spaces.[26] The presence of female migrants introduced South Asian customs of ritual gift-giving (*lena-dena*).[27] Migrant women would organise social events in their homes, where the exchange of gifts, and the subsequent sequences of reciprocity, would help engender relations between families. With the advent of the second generation, further ties were forged through the arrangement of agnate marriages. These developments led to the gradual consolidation of novel communities, united by a common culture and migrant status.

> Before I first came to England, I was very excited. People from my village who were already there, used to write letters back home. And I remember people talking about how the Pakistanis used to work side-by-side with White people. It seemed strange to me, because I never met a White person. So, when it was my time to go to England, I was very excited. I thought it would be a good experience working with English people, and seeing how they lived. I wanted to make White friends and play cricket with them. Back then, I was a pretty good spinner [*laughs*].
>
> But when I came here, everything was different to what I expected. The Asians were working on different shifts to the Whites. We usually worked the night shift at the factory, and it was very difficult to meet White people [...] Other White people in the town were very distant, and many of them used to call us racist names

in the street [...] Those who didn't, felt hatred inside – you could just sense it. In the end, I decided it was better to stick to my own. Pakistanis won't hate you because of the way you look, because they look the same! We used to live together, work together, play cards and cricket together on our day off. That was our lives. Of course, these are the days before the wives arrived, so no one was there to tell you what to do! [*laughs*].

– Malik, 64

***

The majority of Muslim households in Luton were patriarchal hierarchies. The patriarch – most often a first- or second-generation migrant – commanded absolute respect and obedience from all others in the household, at least in theory.[28] He controlled family finances, had the final say on wedding arrangements and funerals, and arbitrated any disputes between family members. When a patriarch died or became incapacitated, he was usually succeeded by his eldest son. The women in the household were subservient to the men, except when their age or rank within the family was above a man's. For example, the wife of the patriarch had less 'power' to make key decisions than her husband. However, she outranked any sons while her husband was alive. Similarly, an older sister outranked all younger siblings, including any brothers. Despite this, in practice, once a man matured, his decisions and choices often trumped those of any female peers and even his mother's in most cases. Many of my male informants, young and old, claimed that women in their household were their 'honour' (*izzat*), and that they should be 'protected'. Furthermore, rules that applied to them did not apply to their sisters. Omar's insistence that his sister should not be allowed to have a boyfriend, for instance, was typical.

Perhaps the most striking manifestation of patriarchy within the Pakistani diaspora was the institution of the *biradari*[29] (literally 'brotherhood', for which read: extended kinship group or clan), which remains functional and avidly observed in Luton even today. Again a throwback to rural South Asia,[30] clan solidarities were mobilised by pioneer migrants when sponsoring kinsmen in moving to Britain through the process of 'chain migration'.[31] Once in Britain, new arrivals would settle in the vicinity of pioneer migrants, who would be a source of support in those early years. Wards and neighbourhoods in migrant areas became steadily populated by members of the same clan or village.[32] Over time, this system of patronage developed into a social network, where clan members and fellow villagers would visit each other in their homes and congregate on special occasions, such as *Eid*.[33] The importance of *biradari* networks among the Pakistani community (and to a lesser extent, the Bangladeshi community) in Luton cannot be overestimated. Through *biradari* networks – controlled exclusively by men – weddings are endorsed and arranged, social and cultural spaces are created, business partnership are established, mosques are funded and built, and political alliances and conflicts are enacted.

> Individuals are subordinated by its demands and its economy – social, moral, and cultural – which operates through the regulatory mechanisms of honour and shame, policing their behaviour, and inhibiting actions that might threaten cohesion and self-identity of the clan. (Mondal 2008:143)

The various actors and players within the *biradari* dictate the fortunes and reputation of the clan. Throughout their formative years, young Muslims become well acquainted with them. They also become familiarised with the structure and mechanisms of their respective *biradari*, in addition (to a lesser degree) to those of others.[34]

A major space within which *biradari* solidarities were affirmed or denied is the local mosque. Luton had twenty-five mosques of different denominations, mostly situated in the Bury Park area. A significant number of these mosques were owned and run by clan groups. The mosque was a place for Muslims, particularly men, to meet and socialise in addition to observing prayers and religious ceremonies. Muslim children (boys and girls) also regularly attended the mosque to learn to read the Qur'an in Arabic and how to perform the daily prayers. The mosque also acted as an arena for local-level politics between different clans vying to maintain or take control of the mosque committee or to assert their community leadership credentials to the congregation.[35] Young Muslims were expected to behave according to religious and cultural tenets when in the mosque environment. They were also expected to attend the mosque that their respective *biradari* supported. Attending other mosques potentially damages the honour (*izzat*) of the clan. Building and controlling a mosque was the source of great social capital, and crystallised the honour of a given clan. It was interpreted as a spiritually and socially benevolent deed. Thus the mosque was where young Muslims received their earliest religious instruction, a place where they continued to frequent throughout their lives to perform prayers, to consult the imam, as well as to attend religious festivals and ceremonies. It was also a place where the elders of the community met and socialised, and asserted their personal and clan interests and ambitions to the congregation. Young Muslims were expected to observe the protocols of the mosque with diligence as it was seen, first and foremost, as a place of spiritual reflection and religious instruction. To many of my young informants, however, it was also the arena where the honour and prestige of the *biradari* was reified.

In addition to the mosque, the local neighbourhood was an important stage where community relations are played out. Some wards and neighbourhoods in Bury Park were almost exclusively populated by Muslim households. Local businesses were also often owned and managed by local Muslims. The close-knit nature of the community meant that young Muslims were aware of the gaze of community elders and family members when out and about in the neighbourhood. A trip to the local shop may entail the passing of an elder from the mosque on the way and the exchanging of salutations (*salams*). Once at the store, the clerk at the desk may be another family friend who is addressed as 'aunt' (*khala*). On the way back, older cousins and their friends may be loitering in the local park. Displays of respect and appropriate etiquette, therefore, were constantly observed in the local area. A youth cannot be seen drinking alcohol, wearing immodest clothes, or flirting with the opposite sex, for example. Such behaviour would bring shame and embarrassment to their respective families, and lower their social standing. Throughout their formative years, Muslims in Bury Park were encouraged to internalise the role and significance of the family and local community. Moreover, they were further expected to embody the status and reputation of their family and clan, and add to it through their good manners, deeds and achievements.

***

Pnina Werbner notes that 'the very notion of *izzat* [honour] connotes, ambiguously, status, rank, honour and dignity' (2002:19). For first-generation migrants in Luton, preservation of *izzat* was an imperative. *Izzat* could be attributed to a particular individual or collective. Religious piety, altruism, wealth and philanthropy, caste and education were all factors that contributed towards the consolidation of *izzat*. Those who were lacking in these respects brought shame (*sharam/besti*) not just on themselves, but to their families. If a family developed a bad reputation, they may suffer stigma in the community. Business pursuits may be disrupted, potential marriages may be annulled, and social exclusion may be enforced. The onus was thus upon all the members of the family and/or clan to uphold its honour. Liaqat, 57, explained to me what *izzat* meant to him:

> *Izzat* is very important. It makes you who you are. People will look at you in your community and think good thoughts of you, if you are from a good family. I always say it doesn't matter how wealthy you are, or what caste you are from, if you don't act in a good way. That's what *izzat* means to me. You must always be honest and worship God. Your children and family members must be good people. If they are not, what's the point [of living]? Not only will you be hated by [other] people, but also by God.

For Liaqat, *izzat* represented a code of behaviour which was heavily associated with religious morality. He dismissed wealth and other worldly aspirations in favour of being a 'good' person, usually represented by good moral conduct. Omar also associated *izzat* with good behaviour:

> You don't wanna bring shame on the family [because] you'll get murdered! My old man's always banging on about it these days, as I'm getting older and know what it means [now]. They want you to act in a way that doesn't embarrass them. Y'know, don't get hooked on drugs, don't fuck around, go to the mosque, that sort of thing. It is good, because it means you won't go astray like other people [in the neighbourhood], but it's hard [because] there's a lot of temptation around. I'm not that good at it, to be honest. I do try my best, [but] sometimes I fuck up. Have a spliff, miss prayers. Actually, I do that a lot! [*laughs*]. But I definitely see the need for it. It makes us better, doesn't it? Although sometimes people take it too far and starts fights over it – if someone insults someone – happens all the time with Pakis, even at the mosque. 'You've insulted my *izzat*'; 'you've done my *besti*'. That's when it gets silly.

Other young informants were also quick to point out that the *biradari* system is easily manipulated by the nefarious intentions of some of its patriarchs. Amin was a twenty-three-year-old history teacher who lived at home with his parents,[36] along with his older sister and brother. Like Liaqat and Omar, his family were originally from Pakistan. Amin's father was a taxi driver, while his mother worked as a dental nurse. Like Amin, his sister also went to university and trained as a television producer. His older brother worked as a delivery driver at a well-known pizza chain. Amin was very critical of the way 'honour' and clan status was cynically appropriated by some within the community to suit their own ends:

> *Biradari* isn't there so you can take advantage of people, look down on them, or use your status for your own goals. A lot of people do that. They abuse its meaning. To me, a family is good if they don't do that. I've seen elders start feuds between other rival families so they can get into the mosque committee and become powerful in the eyes of the community. Or use the fact that they're rich, and effectively bribe people to fuck up business deals so no one can compete with them. No one says anything because they've all been paid off, or have their own interests. The women are just as bad. They go around spreading rumours about other women – about their character. This is so they can't get married. Purely out of jealousy. [And] because they are from good *biradaris*, people listen to them. This is bullshit. I'm sure *izzat* isn't supposed to mean this.

Amin's frustration with clan-based rivalries and intrigue reflected the attitudes of a number of my older informants, especially those I met through the Salafi Mosque and at the offices of MSP. Many claimed that clan politics created disputes, divisions and tensions (*fitna*) in the community and argued that it was forbidden (*haram*) in Islam to hold on to such loyalties. As a member of the second generation, Khidr was very familiar with *biradari* politics and very critical of it. He argued that loyalties towards one's clan over other Muslims was a throwback to the ignorance (*jahaliyya*) of pre-Islamic Arabia. He referred to it as '*asabiyya*', an Arabic term roughly translated as tribalism. According to Khidr, such tribalism was a symptom of pre-Islamic decadence, and the purpose of Mohammed's mission was to challenge such inharmonious doctrines:

> People should be loyal to God, not to their blood. Muslims should be united, as brothers and sisters. We are one. Unfortunately, a lot of the older generation still hold on to these backwards concepts, a lot of which comes from the village [in South Asia].

Khidr went on to describe the problems from his own extended family when he decided to marry his first wife. She was of Caribbean origin and had converted to Islam. Even though she was a 'good, pious Muslim' who wore the headscarf (*hijab*) and thobe (*jilbab*) and observed correct Islamic manners (*akhlaq*), many members of his *biradari* disapproved of the marriage. Fortunately for Khidr, his own parents were very supportive and so the marriage ultimately went ahead. Khidr's parents were impressed with the fact that his prospective wife was a good Muslim, despite being from a non-Muslim background. However, this attitude remains quite rare among the older generation. The majority of my informants claimed that their parents and/or grandparents (depending on their age) still maintained preference for consanguineous marriages within the *biradari* or caste group.[37]

***

Marriages among cousins was still very common among the Pakistani community in Luton, with many of my younger informants either anticipating it in the future, or already preparing resistance against it. The practice had its roots in rural Mirpur, where a combination of factors, such as the retention of wealth within the kin-group, facilitated by Islamic laws on endogamy and inheritance, and consequent intensification of sibling ties, had created a preference to marry 'within'. In addition to the practice's rural roots, mass migrations from Mirpur in the twentieth century, and the considerable increase in wealth that came with it, further consolidated the practice in diaspora, especially in the early phase of settlement.

> Although the majority of British Mirpuri marriages are still *biraderi*-endogamous, it does not follow that all such marriages involve Pakistan-based partners; many are arranged between British-resident couples. Nevertheless, although many younger British Mirpuris have now grown [...] doubtful about the viability of matches with spouses from back home [...], their parents often find their hands are much more closely tied, feeling duty-bound to consider offers of *riste* [marriage proposals] from Pakistan-based *biraderi* members just as seriously as those proposed by local residents. (Ballard 1990:243)

Over two decades on since Ballard's observations, in Luton at least, the pressure on young people to marry cousins remained seemingly acute. Many young third-generation Muslims expressed grievances when the issue of arranged or cousin-marriages was raised. The primary concern was a lack of choice. Despite this, they were all aware of the importance of marriage in maintaining *biradari* ties, and great effort was still made by parents in trying to preserve this central institution.[38] Accordingly, any potential or proposed marriage must be endorsed by *biradari* members in order to avoid stigma or marginalisation by the family. Whereas, in the early phases of settlement, the stigma may have been overbearing, my informants were finding ways of negotiating with their parents, particularly utilising reformist religion as a means to counteract 'cultural' preferences in marriage. Shahed was a nineteen-year-old economics student whom I initially met in a prayer room at a prominent London university. Shahed's family were members of one of the largest *biraderis* in Luton and he had many female cousins there who were of marriageable age. Being academically gifted, pious and of gentle character, he was considered to be a highly eligible bachelor. His eligibility was heightened when many of his male cousins dropped out of school early and were involved in petty criminality. Reluctant to marry a cousin, Shahed was nevertheless resigned to the notion that, ultimately, the decision was 'out of his hands':

> Given a choice, I want to marry someone outside the *biraderi*. I have to be able to get along with her and share the same interests. I don't think any of my cousins fit that role, and I don't really know the ones in Pakistan. But in the end, I know I'm going to have to go with whoever my parents like […] I'm just going to have to say no until I like someone they like. But I'm not really thinking about that now. I want to finish my degree and get a job. Maybe then the situation will be different, who knows?

Shahed's stance was unusual when compared to many of his peers. He was willing to bend to the wishes of his *biraderi*, whereas others were more critical. An increasing trend among such critics was a countering of the 'village mentality' of the elders with 'pure Islam'. They were quick to point out that, according to Islamic law, a man could marry anyone so long as they were Muslim. Those who were a little more learned in theological matters, even argued that Muslim men were allowed to marry Jewish or Christian women who were 'chaste', as they were considered 'People of the Book' (*Ahl al-Kitab*). (Although the same was not the case for Muslim women.) This fact was an 'ace up the sleeve' for the majority of my informants and many intended to pull it out at the right time as a bargaining tool. If Islam allowed it, how can the elders refuse? Especially since they were led to believe that the doctrine of *izzat* was synonymous with religious morality and etiquette? Omar, for example, categorically rejected any prospect of an arranged marriage, let alone marrying a relative. However, he was far more open to marrying a Muslim:

I don't want an arranged marriage. I think they're stupid. First of all, I don't even like any of my cousins. And even if I did, they're too ugly. I'd rather marry someone I get on with, and find her myself. My parents don't have a clue what I'm like, how can they find me someone? […] I don't mind marrying a Muslim, but I want to choose who it is, not my parents. As long as she's Muslim, it shouldn't matter to my parents, but I know they want me to marry one of my cousins – it's not happening though!

Even though the rebellious Omar was resistant to an arranged marriage to one of his cousins, curiously, he maintained a preference for marrying a Muslim, despite having a White girlfriend. When I asked him whether he would consider marrying his girlfriend or a non-Muslim in the future, he said it 'wouldn't work' with his family. He argued that she would find it hard to settle in or accept the customs at home. He claimed that it would just be easier to marry a Muslim girl as she'll be able to understand his family, and get on with his mum; 'at the end of the day, I'm Muslim'.

The idea that a non-Muslim was unable to 'fit in' to the family was a pervasive one for many of my informants. While the majority expressed a desire to 'marry out', or have the final say in the marriage, a few consciously opted to marry a cousin from Pakistan of their family's choosing. Of these, all were second generation. One such informant claimed that British girls had 'corrupted' moral values, whereas a 'village girl' in Pakistan was typically 'innocent' and, furthermore, because she was related to him, would have a vested interest in looking after the family. Such attitudes, however, were admittedly rare among my informants. The majority acknowledged that it would be difficult to assimilate non-Muslim partners into the family structure. Kaiser was a twenty-one-year-old man whom I met at the offices of the MSP. He was unemployed, and had been moonlighting as a cannabis dealer to supplement his income. Kaiser had been involved with an older White woman since he was seventeen and claimed the relationship was quite serious. So much so that she had converted to Islam in the hope that they could get married without any issues. In spite of this, Kaiser was pessimistic about the future. He couldn't imagine any circumstance where his wider family would accept the marriage, as they were quite 'traditional' and set in their ways:

I know my family will never accept her because she's White and not from my biradari. So I haven't told them about her. They've lined me up with a girl from back home and I can't really get out of it because my whole family will switch on me, and not speak to me and shit. If that's the way they want it, then I'll marry her and carry on seeing my girlfriend. I've spoken to my missus about it and she understands. At the end of the day, my mum's really important to me and I know my dad will stop me from seeing her if I don't get married back home. So I'll do what they want me to do, and [then] carry on with my life.

Kaiser's insistence on continuing with his relationship even after he's married another was consistent with attitudes held by others, particularly among some older second-generation men. According to Khidr and Ammar, there were a number of men in Luton, many of them their childhood friends, who had married in Pakistan due to family pressure but simultaneously continued with their previous relationship. Some even had children from both partners, and juggled their time between the two households. In most cases, the first partner knew of the second, however, the latter or her family had no idea, at least initially. Once again, some justified these marriages through the prism of Islamic law, arguing it was religiously acceptable for men to engage in polygamous marriages. In fact, some were associates of the MSP through the Salafi mission, and were engaged in the *dawa* (preaching) circuit. For them, religious piety offered a socially acceptable way out of the 'culturalist' confines of the *biradari* structure. For my young informants, however, only time will tell how the issue of marriages pans out for them. In the buoyant words of Shahed, when the time for marriages comes, 'maybe the situation will be different, who knows?'

Broadly speaking, young Muslim men in Luton were aware of their responsibilities to their family and clan. They had been socialised within an environment where loyalty and solidarity towards the family was a given. This included embracing certain overarching customs, attitudes and patterns of behaviour. The home and local community, (which also extended to their ancestral villages in South Asia), was the arena where such relations were enacted. For many, this was a necessary part of life, and brought a number of benefits. As we shall see in the following chapter, being bonded to the family, and its extended social networks, reaped material and emotional benefits. It remained, in some cases, the basis for political solidarity against antagonisms from beyond the community, both real and perceived – as it had also been for the pioneer generation. On the other hand, many of my informants found some aspects of the bondage to be oppressive, not least in the spheres of maintaining 'backward' village, caste and clan rivalries from 'back home', which many had deemed obsolete. By far the most contentious issue, however, was marriage. Young men (and indeed, the majority of women that I encountered), sought a choice in the matter. Some were altogether contemptuous. Of those, religion was utilised to counter the normative structures of

South Asian (read: 'un-Islamic') 'culture', rendering them inappropriate for the rigours of modern life in diaspora. This 'tactic'[39] is both compelling and increasingly efficacious. 'You can't argue with God', Khidr reminded me. 'The elders are slowly coming round to that fact'. For now, at least, they remain seemingly intransigent.

Notes

1. See Franceschelli 2016; Becher 2008; Tarlo 2010; and Scourfield *et al* 2013.
2. See Anwar 1979; Ballard 1988; Gardner and Shukur 1994; and Gardner 1995.
3. See Adams 1987; Cressey 2006; and Mondal 2008.
4. See Shaw 1988; and Werbner 2002.
5. See Patterson 1969; and Solomos 1993.
6. See Dahya 1973; Jeffery 1976; and Anwar 1979.
7. See Ballard 1994; and Gardner 1995.
8. See Werbner 1979, 1980; Adams 1987; Eade 1989; Shaw 1988; and Ansari 2004.
9. See Anwar 1979.
10. Compare with Elgar 1960; Gardner 1995; and Lyon 2004.
11. See Allen 1971: 66.
12. Compare with Mines 1994.
13. See Ballard 1990:229.
14. See Anwar 1976; and Watson 1977.
15. See Shaw 1988:111–33; and Werbner 2002:185–210.
16. See Jacobson 1998.
17. See Modood and Werbner 1997.
18. Compare with Lewis 2007.
19. See Roy 2004.
20. See Ewing 1990; Werbner 1996; Bayart 2007; cf. Marsden 2005; Schielke 2009; and Simon 2009.
21. See Adams 1987.
22. Kotli is a large town and region in Mirpur District, Pakistan. The vast majority of my informants of Pakistani heritage in Luton originated from this region.
23. See Foreman 1989:26–8.
24. See Wemyss 2006.
25. See Werbner 1990:50–78.
26. See Eade 1996.
27. See Shaw 1988:112.
28. Compare with Archer 2003; and Simpson 2013.
29. Most British-Bangladeshi use the term 'gushti' to refer to the patrilineage. Although there are some marked similarities with the biradari model, particularly in the process of chain migration and early settlement, the gushti plays a less prominent role in social, economic and political spheres in diaspora when compared to British-Pakistanis.
30. See Blunt 1931; Elgar 1960; and Alavi 1972.
31. See Lewis 1994; Werbner 2002; and Cressey 2006.
32. See Eade 1989.
33. See Shaw 1988; and Werbner 1990.
34. Compare with Din 2006.
35. Compare with Raza 1991; and Werbner 2002.
36. Amin lived and grew up in the nearby town of Bedford. He spent a lot of time in Luton, visiting friends, and competing in the various football leagues as a goalkeeper.
37. Compare with Ballard 1990; Modood and Werbner (eds) 1997; Shaw 2001; and Hussain 2008:130.
38. Compare with Shaw 1988:85.
39. See De Certeau 1984; compare with Fadil 2009; and DeHanas 2016.

3
Friends

The previous chapter showed how the early settlers in Luton tried to carve out a community space when faced with the prospect of settling permanently in Britain. It charted the way in which the migrants grappled with pressures to provide for their families while, at the same time, managing the relative alienation that came with moving to a foreign country at a time when economic outputs were low and racial tensions were high. We also saw the way in which 'chain migration' and 'white flight' shaped the physical contours of the neighbourhood, and established communities clearly stratified along the lines of race and class. Recognising the nascent sense of permanence, early migrants set about adjusting to life in Britain for the long run. Children were sent to school, homes were acquired and a whole host of new commercial enterprises took root. Relationships with fellow migrant kin and villagers were consolidated, and effort was made to forge lasting social networks of ritual exchange and reciprocity in order to ease the settlement process. Within the home, the second generation were socialised in a curious way. They were expected to observe and maintain certain cultural practices that developed in rural areas of South Asia, superimposed by their parents in an effort to maintain their 'roots'.

These practices and attitudes contrasted with those that were pervasive beyond the boundaries of the community. At times, they conflicted. The offspring of early migrants, therefore, attempted to tactically navigate these often contradictory moral registers in their everyday lives. Some eventually succumbed to parental and extended family pressures and accepted consanguineous marriages, for example. Others rejected such pressures outright, and faced the wrath of the community by being shunned and stigmatised. Others still, tactically appropriated religious law as a way of countering the unreasonable and obsolete 'village mentality' of their parents and wider clan. Despite these pressures, young Muslims were, naturally, exposed to 'others' from different ethnic and religious backgrounds from an early age. In addition to those influences from home, these interactions had profoundly shaped the way they self-identified and how they viewed others in society. In many cases, young Muslims had an array of non-Muslims who are regarded as friends, mentors, colleagues, teammates and even lovers.

This chapter attempts to explore such relationships and their role in shaping unique identities that have emerged through the experience of living as a minority with complex loyalties and moral outlooks. It will investigate the role of social environments and spaces beyond the 'home' – such as school, university, sports and social clubs, holidays abroad and the workplace – in making friends and socialising. Curiously, young Muslims did share stark similarities in attitudes with their elders when it came to earning a living and contributing to the family. They all recognised the need, as 'men', to earn as much as possible so that they could 'provide'. Some chose to go to university, in order to command a higher salary upon graduation. Others went straight into work after finishing school. Some even ventured into crime in order to 'give something back' to their parents. This chapter will thus also explore the role of work in the lives of young men in Luton – what motivates them to work? What kind of work? And, when they are not working, how do they spend their free time?

***

Ali was nineteen and academically gifted. His parents (from the second generation) were followers of the Salafi creed, and were regular attendees at the mission. He was the youngest of three and the only son. Ali's parents never made it to university and, consequently, there was some pressure on him to follow his sisters and win a place at a reputable one. Although his parents' preference was for him to study medicine, he hadn't been offered a place. At the time of fieldwork, he was sitting two further A-levels in order to boost his chances. Like many young men of his age in Luton, Ali was a keen sportsman and played for a local cricket club. Unlike many of his teammates, however, he was an observant Salafi – or so his parents thought. On the surface, Ali prayed five times a day, fasted during Ramadan and abstained from mixing with the opposite sex. When with his friends, however, he was less stringent. Ali claimed that he 'practised' Islam because he believed in its message but also because his parents insisted. He also went on to concede that there were many moments of slippage.

Salafism is an austere and puritanical form of Islam that requires meticulous attention to personal codes of behaviour and observance of daily rituals (of which there are many). The doctrine is bound by a form of conservatism that strictly forbids any imposition of modern innovation (*bid 'a*) into religious thinking or practice.[1] Only the content of the Holy Scriptures (i.e. the Qur'an and the Tradition of the Prophet) are to be taken as sources of human inspiration. Any authority beyond that is rejected as heresy and the malevolent murmurings (*waswasa*) of the devil (*Shaytaan*). Salafis also reject all other sects and schools of law and theology in Islam. They argue that all of them have skewed the True Path to God throughout history, and introduced a host of non-Islamic innovations to the religion, including doctrines of rationalism, mysticism and the 'cult of jurists'. Rather, they hold that the True Path lies in the strict, literal application of the Text in everyday life. Further, they argue that this application requires no further conduits (such as a cleric or saint) for a lay Muslim, other than the ability to engage with the Text directly. Knowledge of the Qur'an and the Tradition, therefore, is highly prized within the Salafi movement, somewhat ironically. Without knowledge of classical Arabic (*fusaha*), for example, a lay Muslim is unable to directly access the scriptures, as they are written exclusively in that language. In such circumstances, any 'believer' in the Salafi method (*manhaj*) is compelled to consult someone who possesses such knowledge and training (*sheikh*). Ali, like his parents, had not received such training. Thus they relied on the interpretation of learned individuals such as Khidr for religious instruction (although Khidr would deny that he is, in any way, an 'interpreter'). Due to this approach to Islamic knowledge, Salafis refer to themselves as the 'saved sect', but under the rubric of the 'People of the Tradition and Community of Believers' (*Ahl al-Sunnah wa al'Jama'ah*) shared by Sunni Muslims more broadly. Salafis take their name from their proclaimed exclusive following of the 'Pious Predecessors' (*as-Salaf us-Salih*) who lived during and shortly after the time of the Prophet (and therefore held 'true' knowledge of how Islam should be practised according to Prophetic Tradition/Example). Ali was only too aware of this founding creed (*aqidah*), having been introduced to it at an early age. Moreover, Salafis (whether lay or learned) made a conscious effort to cyclically reaffirm this religious identity in their everyday rituals and practices.

Although Ali claimed to be a genuine member of the sect he was also aware of his laxities, especially when with friends. He didn't admit to having a girlfriend to me, but did concede that many of his closest friends were not 'practising', and smoked weed, drunk alcohol and partied. When I asked him whether he joined in with the mischief, he responded with a wry smile: 'I just chill with them'. Ali was clearly torn between the abstinent demands of his committed parents and being a young man in twenty-first century Britain. Unlike Ali, however, the majority of my young male informants were not from a 'Salafi family'. Comparatively, the Salafi community in Luton remained quite small, although numbers are rising (a point I shall return to in the final chapter). The majority of Muslims in Luton followed the Sunni Hanafi school of law,[2] which is the dominant strand of Islam in South Asia.[3] Most of Ali's friends, therefore, were not Salafi. Some were non-Muslims, including those of Sikh, Black Caribbean, and White British heritage. The fact that his friends included those who were not practising Muslims, or Muslims at all, was a point of concern for Ali, as his parents had always encouraged him to seek 'pious' friends. Having such friends equated to good company and thus good influences. Of course, being 'pious' meant being Salafi. Although Ali did have some friends from the mosque, he preferred spending time with his 'actual' friends. In this Ali was not alone and many others had similar dilemmas. Although their families were not always Salafis, like Ali's, they too were nevertheless burdened with comparable morally conservative demands from home. For many, being economically active and providing an income to the family was prioritised,[4] and formed the basis of an essentialised masculine sensibility.[5] This meant getting a job as soon as one could. Or, if academically competent like Ali, attaining employment on completion of further studies. Once economically viable, marriage and settling down was the next step. All the while, one was expected to preserve the 'honour' (*izzat*) of the family and to actively promote its status and prestige in the community. This included acting with decency and in accordance to established protocol when in the public eye; abstaining from all forms of proscribed excesses, such as drinking and gambling; and totally limiting all forms of sexual or criminal scandal. Naturally, reality was some considerable distance from the ideal. All of my informants struggled to put these expectations into practice.[6] Some didn't even bother.

A key chasm between early migrants and their British-born descendants was the level of exposure to wider society. Early migrants were pressed for time and were chiefly concerned with earning as much money as possible before their time 'ran out'. As we have seen, this never actually happened and the majority remained in Britain and settled their families here. Despite this, working hours, ethnically homogenous neighbourhoods and social networks, and a yearning to continue cultural ties with South Asia, meant that a process of collective 'insularity' took root. Exposure to non-migrants was limited to colleagues at work (if any), when liaising with local authorities and the state, or other quotidian encounters beyond the confines of the community. The second and subsequent generations, however, were schooled and socialised in Britain and so, in comparison to their parents and grandparents, were far more exposed to non-community members in their everyday lives. Crucially, they had more time for leisure and recreation. More time to access the arts, the media, to shop, to play and to make new friends. In essence, to have 'fun'.[7]

> When my family came, I had many new responsibilities on top of my job. I had to find somewhere to live. I had to find places in schools for my children. Take them to the doctor. Everything changed, y'know? My wife needed help with the shopping, so I used to spend my day off going to the market and doing things around the house. You didn't have time to do anything apart from working and taking care of your family. My friends all lived near me, but I hardly ever saw them. Every now and again you would see them in the mosque, or invite them round for tea. Apart from big occasions such as Eid, or at weddings, you never really had the time to see them. But, in England, that's how life is.
>
> – Malik, 64

Although young Muslim men in Luton generally worked long hours, the overall time that they spent at work was still considerably less than their forebears. Moreover, the types of jobs that they were engaged in were far less gruelling in comparison, and paid considerably more. This is due to the fact that the range of jobs available to them was more diverse, as a consequence of their formal education, language proficiency, or specialised skill-sets. Furthermore, unlike early migrants, young Muslim men did not have to reconcile the pressures of being an integral financial provider for a large number of dependents, both in Britain and South Asia. Thus, they were in a more privileged position than their predecessors and could enjoy a richer and more varied social life. Those who pursued higher education were in a unique position where they would leave the community for a few years, make new friends and acquaintances, and return home wielding a more elevated social status, and harbouring fresh tastes and perspectives. Young men such as Omar and Ali, who were neither working nor attending university, spent a lot of their time socialising with school and college friends, playing sports, and planning mischievous 'nights out', in the firm knowledge that it was a fleeting period in their lives as men and as Muslims. 'I can't always be this way', Omar said to me 'One day, I'll have to be a good Muslim'.

***

Muslim men of all ages in Luton demonstrated an indefatigable will to be economically active – an attitude that remained heavily informed by the memory of migration. On the whole, Muslim men in Luton tended to be employed as low-skilled manual workers in factories and warehouses, or are petty entrepreneurs. Most commonly, they were found in the retail, wholesale or catering sectors. Driving taxis was also popular among the first and second generation, due to its flexibility in choosing hours and the relatively decent pay. In recent years, and particularly among the younger generation, a trend has developed where employment has also been sought as private security officers. The long working hours, good pay and more reticent nature of the job appeared attractive to many. Interestingly, some also claimed that, given the choice, they would prefer to work for 'Whites' (*goreh*). By this, they meant working in an environment where wages are regular and paid on time. Faheem, a third-generation Bangladeshi, worked for a large electrical retail company in Luton as an assistant store manager. He left school at sixteen and immediately ventured in to the world of work. He described his decision not to go on to the sixth form as 'the best [he's] made' because, aged twenty-three, he was financially better off than some of his 'pathetic' friends, who were unemployed having returned from university:

> I'm always teasing [them] that they wasted years of their lives when they should've been working. I feel sorry for them as many of them either don't have a job, or are just starting out at jobs they could've been doing years ago.
>
> I've been working now for seven years, and I have saved quite a lot of money in that time. When my fourth sister got married recently, I was in a position where I could help my dad and contribute to the wedding. I've also managed to buy some properties over the years, and now I'm financially secure at quite a young age.
>
> My friends should be in the same position – as Asians, you live at home and don't really have many outgoing costs. You can save a lot of money and continue the work of your parents. To me, it doesn't make sense to go to university if you're not going to become a doctor or lawyer.

When I asked him why the usual allure of fun and adventure didn't sway him to go to university, Faheem responded that he was always pushed to make something of his life by following in the footsteps of his father, paternal uncle, and paternal grandfather – the pioneer migrant. The latter had been a successful businessman and landlord, who owned a number of Indian restaurants and residential properties. Faheem cited the old migrant adage to me, that his grandfather had come to England from Bangladesh with 'nothing' and, at the time of his death, was a very wealthy man. After his death, Faheem's father (the elder brother of two) and his uncle took over the businesses and ran them together. Faheem, however, chose not to go into the family business due to the anti-social hours. He also wanted to do something different. Despite this, he insisted that his work ethic and desire to expand the family's combined portfolio were driven into him at an early age. He was very proud of his family's reputation in the community as 'hard workers' who did well. It seemed important for him to maintain that image. Even if it meant not going to university and 'wasting' valuable time: 'if you're Asian in this country, you have to work. That's why we came here in the first place, and that's the way it should always be'.

Riaz was nineteen and worked for his family at their grocery store on Dunstable Road, the main thoroughfare in Bury Park. Like Faheem, he left school at sixteen and decided to go straight into work. However, unlike Faheem, Riaz's decision wasn't solely based on his desire to contribute to the family, but on how they perceived him:

> In my family, you're expected to work for the family when you get older. You can't sit at home and get everything provided for you – you have to earn it. A few of my friends don't work, and they get a lot of bother at home from their parents [for it]. My parents think they're a bad influence.
>
> I don't want to let down my family. I don't want them to look bad in the community because of me. It doesn't go down too well if you have people in the family that can work [but] choose not to. Especially when you have a family business.

Riaz had a strenuous working schedule, which was typical of many Muslim men in Luton. Interestingly, he provided his labour to the family without demanding a formal wage. He worked six days a week at the shop, starting at six in the morning and usually not finishing until seven or eight in the evening. His family did provide him with some spending money for his efforts. The amount varied from week to week:

> I live at home, so I get off paying for rent and food. To be honest, although I feel like I'm always working, it's alright – at the end of

the day, it's my business. I'll get a share in the end, so it's okay – family always takes care of you.

Nizam was in a similar predicament. At twenty-five, he was a few years older than Faheem and had more responsibility. He too started working for his family straight after he left school. The family ran a wholesale fruit and vegetable company, supplying to small businesses all across London and the Midlands. All the male members of Nizam's paternal family were involved in the business. Unlike his brothers, however, Nizam was bookish. He always harboured ambitions to go to university and read history. At the time of research, Nizam had been accepted at the London School of Economics to attend a short introductory course in Modern European History. Like Faheem and Riaz, Nizam chose to work for his family because it made his father happy:

> For me, the family is the most important thing. I'll do anything for them [...] My old man wanted me to help him out in business after I finished school, as he was getting old [...] He wanted me and my brothers to take over. So I didn't think twice.
>
> Almost ten years on now, I'm still doing the same thing; getting up at four in the morning to drive the van down to Spitalfields; breaking my back grafting in the freezing cold; spending the day delivering all the goods. It's a hard life, but it makes my old man happy.

Nizam's father was the chief patriarch of his clan (*biradari*). Back in Pakistan, his father, with the assistance of relatives based there, had built a business empire in commercial and residential property development. In England, Nizam's family owned and rented out a number of residential properties, in addition to their wholesale business. Despite this success, his father remained ambitious, and insisted that his three sons expand the business in his absence. His two sisters have been married off to cousins (of his father's choice) in Pakistan and, shortly after the conclusion of fieldwork, Nizam himself married a British-Pakistani cousin, despite being in a long-term relationship since he left school. Although he claimed to be very much 'in love' with his girlfriend, in the end, he chose to honour his family and grant them their 'wish', as ever. By 'family' however, Nizam seemed to be referring exclusively to his father. It appeared to me that his mother was conspicuously absent from all formal decisions pertaining to the 'family', despite Nizam's very close relationship with her. When I asked him why this was the case, he replied that women in his family never make such decisions and that these were firmly in the remit of 'men'. In Nizam's family, like so many other Muslim households in Luton, there was a division of labour along gendered lines. Men made the key decisions and were the chief breadwinners, whereas women took charge of all domestic work – cooking, cleaning, shopping, hosting guests as well as remaining hidden from gaze (*purdah*) and being obedient to the men. When unmarried men or women reached marriageable age (usually eighteen-plus), they were matched up with a suitable partner of the patriarch's choosing. Of course, over time, the power of the patriarch in the British context has been waning. Young Muslims have been registering resistance to such polities, which are seen in the main as antiquated and inappropriate for life in Britain. However, in Nizam's family's case, progress had been less forthcoming.

Nizam graduated from secondary school with ten GCSEs,[8] five of them at grade A*, the top grade. His teachers had always encouraged him to further his academic prowess by going on to university and attaining a degree. Although he was interested in doing so, Nizam was aware of his responsibilities at home. I asked whether it would ever be possible for him to return to his studies and he remained, on the whole, optimistic:

> Well, you never know, do you? At the moment, I'm doing this course, which I really enjoy. I spoke to my dad and [older] brother about starting a course in September, but they think it's not a good idea.
>
> My dad doesn't see the point of me starting a degree at the age of twenty-five. He says to me: 'why you do you need a degree? You have a good job. You make a lot of money working with your brothers'.
>
> My father is 'old school', so he doesn't agree with the idea of studying to enrich your mind, and neither do my brothers. For them, people should only study to get a better job and earn more money. Obviously, studying history isn't going to put me in that situation is it?

And he makes a pertinent point. Most Muslim men in Luton wanted to make as much money as possible. The method for achieving this was working long hours as well as disciplined thrift. My young informants who were working, all lived at home. This meant that they did not pay for board. Many of them worked for their families, like Riaz and Nizam, which allowed for wealth to be pooled in order for it to be spent on collective ends, whether new business ventures, funding weddings, or building mansions (*koti/basha/rajbari*) in South Asia. The income contribution from young male members of the family was therefore integral. Furthermore, the family network was an important means of providing employment. Members of the community who occupied a considerable position of power and influence within a company were, on occasions, obliged to recruit friends and relatives upon request. Faheem admitted that he had recruited friends and family in the past, some of whom continue to work with him:

> My friends are always asking me for jobs [...] because they know I got a job for a couple of my cousins. I don't mind helping them out, as we look after our own. But sometimes people ask you who have criminal records and drugs problems, and it's really difficult to help them out. I try to do what I can without getting into trouble.

A significant proportion of Luton's young working males, therefore, shared the workspace with members of their family or other Muslim friends and peers. However, some also strayed into illegal work, such as dealing drugs. Importantly, their motivations for doing so remained consistent with the rest. 'Working' and 'making money' were valued status symbols for young Muslim men, just as they had been for their generational predecessors.

***

Among my informants, a number of young men between the ages of sixteen and early twenties, held ideals of becoming a drug dealer. Some even sold drugs on a casual basis, while still at school (under sixteen). According to my colleagues at MSP, consumption and supply of drugs within the Muslim community had rapidly increased in the past ten years. When I inquired as to why this was the case, they explained that Muslim gangs possessed sole control of the drug trade in the town. Previously, they claimed it was 'the Jamaicans'. However, over the course of a series of bitter and violent turf wars in the 1990s and the early part of the 2000s, Muslim gangs eventually triumphed, and came out on top. According to my interlocutors at Bedfordshire Police, gangs controlled the drugs trade in the town and its hinterland. Over time, the groups fragmented into rival factions. During my time in the field, Muslim gangs were embroiled in violent turf wars with each other. One of the earliest comments Fakruddin from the MSP made to me (while he was conducting an impromptu tour of the town on my first day), was that it was the 'drug capital of the Midlands' and, somewhat facetiously, that 'Muslims ran the show'. This information had not been lost on Bury Park's youngsters. Some of the more charismatic leaders of the gangs were venerated as heroes by many of the young men with whom I spent time. A few had assumed mythical status, especially those perceived to have sacrificed their lives or those who were spending lengthy stints in prison. Mukhlis, a Muslim man in his twenties with Pakistani heritage, was an experienced youth worker for Luton Borough Council. He was an expert in mentoring young men with behavioural problems who had a history of petty criminality. He had worked with a number of young Muslim men who had either been, or were later to become, professional drug dealers.

> They grow up idolising these big gangsters they hear stories about [...] who they see around the neighbourhood in their flashy, expensive cars; wearing expensive clothes and jewellery. It's also a bravado thing. Being a gangster means you're tough and powerful. [For them], it's the ultimate sign of manhood. They hear stories about shootings and fights [...] They are impressed by the danger of that kind of life.

As Mukhlis points out, the young drug dealers with whom I struck up a relationship embraced the cosmetic appeals of being a member of an organised gang. However, at the same time, they were keen to point out to me that, ultimately, they were just earning a living and that was the primary attraction. Hadi was a nineteen-year-old man who came from a large family. He suggested to me that beyond earning a living, the 'image' of being a gangster was also an appealing factor in his choice of work. However, he remained convinced that his primary motivation was purely financial:

> Yeah, the whole deal is appealing. Who doesn't like having nice things? [...] Being able to do whatever you want? But, at the end of the day, I got into this because I didn't want to work [...] in a shit job for shit money. [My dad] worked all his life for shit money, and we always had money problems. I don't want my life to be like that. If you're clever, you can make a lot of money from shotting[9] without getting banged up. My two older brothers drive taxis, and work around the clock. I give as much money to the family as they do, so who's laughing?

Hadi had managed to keep his occupation a secret from his family and regularly lied to them regarding his income. His parents were under the impression that he worked part-time as a security guard in Milton Keynes. He did, however, admit that some of the younger members of his family may have worked it out through hearsay or by catching glimpses of him around town. Even so, Hadi was steadfast in his stance of not confirming their suspicions. He was particularly concerned that this may lead to problems with his father with whom he had had a fraught relationship throughout his childhood and adolescence. Hadi and his father were not on speaking terms. This was because Hadi had been charged with shoplifting and other minor crimes in the past. The relationship between him and his mother on the other hand was generally amicable, and it is through her that all his communications with his father were accomplished: 'I get on with my mum, but my dad still hasn't forgiven me [for the past]. He leaves me alone these days, as I give money to my mum, but if he found out [about the dealing], he'll probably kick me out of the house'. Hadi's father had a long-term illness that forced him to retire early. Since his retirement, his older brothers have taken over the role of maintaining the household. Hadi himself had recently started contributing financially to the family. Although his work is illegal and stigmatised in the local community, he vehemently defended his choice:

> Look, I know I'm messed up [...] but I make sure I take care of my family. Yeah, I'm selling drugs, but I give a lot of the money to my parents, so they don't think I'm doing nothing. If I sat around at home, waiting for my family to feed me, it would kill me. Everybody works in my family [...] I don't think I'm doing anything wrong. I don't sell to kids. I sell [drugs] to people who know what they're doing.

Even though involvement in illegal work, such as drug dealing, attracted considerable social stigma for both the individuals involved and their family or clan, this did not seem to deter some of the young men. Hadi deliberately avoided publicly disclosing his profession. This was both in order to prevent the inevitable 'shame' (*besti*) that such news would attract and more importantly, perhaps, to avoid offending his parents. He was also a victim of his past. Previous criminal convictions had meant that Hadi struggled to secure 'legitimate' jobs that he felt suited his talents and aspirations. After a period of unemployment and being financially reliant on his family, Hadi made the decision to sell drugs. Although quite austere in his mannerisms, and often quite melancholic, he struck me as highly intelligent, articulate and, notably, a conscientious person. He possessed a burning desire to provide for his family, especially his mother.

As Fakrudddin pointed out to me early on in my sojourn, the drug trade had a ubiquitous presence in the town. This was especially so in Bury Park, where top-end ostentatious sports cars, laden with young Asian males being driven up and down the thoroughfare, were a common feature of the landscape. It was also common to witness young 'runners' sitting in cars parked up in various side streets, awaiting calls from prospective clients or 'the boss'. Eighteen-year-old Hashim was one of them. A runner was the most junior member of the gang and also its most publicly visible component. He was the street-level conduit between the local supplier and his clientele. His job entailed receiving instructions from the supplier to meet and deliver drugs to clients at a designated location. Runners were issued with a mobile phone and, if they're lucky, a mode of transportation – usually a car or a bicycle. Hashim's supplier was his paternal cousin, and he worked within a wider team consisting of male members from his clan (*biradari*). He had been a runner since leaving school at sixteen, and had ambitions of one day becoming the 'boss': 'I'm happy shotting on the streets [for now], but one day I'm gonna move on to bigger things [...] I work with my cousins [so] they look after me. The higher you go; the more money you make. But I'm doing alright'. Hashim had wanted to be a dealer while still at school, and claimed to have occasionally sold cannabis on school premises. He explained that the transition to full-time dealing was made possible by his cousins paving the way. They were significantly older, had built up connections in the trade and offered him a role as soon as he left school. Hashim was expelled from school before he could sit his final examinations and left without any formal qualifications. At school, he was considered to be a Special Educational Needs (SEN) student, and had a history of violent conduct: 'School was shit. The teachers hated me, and I hated them. I couldn't wait to leave. I knew I could always work with my cousins. I couldn't wait until I started making some money, and I'm doing alright now'.

Hashim's father, a taxi driver, was the head of a household of seven. Hashim had two older brothers and two younger sisters. His two older brothers, who were both married with children, ran a fast-food restaurant together. But Hashim refused to join them, despite pressure from his father. This led to arguments and caused a fracture in the relationship between father and son. Hashim mentioned that his father had apprehended him while he was dealing in the street on many occasions. But he kept carrying on without any further sanctions from home. Like some of the other young men in this study, Hashim also felt the need to contribute to family finances and did so, albeit irregularly. Although his family were well aware of where his money was coming from, they nevertheless accepted his contributions.

> Everyone knows what we do, but they don't complain when the money comes in, do they? Money talks. My dad tells me off all the time, but I just stay out of his way. I give money to my mum, every now and again [and] that keeps them happy.

Even though they were both involved in a profession where income was potentially limitless, both Hadi and Hashim still earnt a relatively modest amount. It is doubtful whether they will ever be able to command the income that they aspired to. The prospect of lengthy bouts in prison was a permanent risk in such a precarious and peripheral industry. The possibility and consequences of physical violence was also a mainstay in their minds. However, despite these risks, both Hadi and Hashim were unwilling to take 'conventional' jobs due to the lower pay and longer hours. By selling drugs, they had discovered a lucrative means by which to earn a living that did not require formal qualifications or clean criminal records, yet provided the scope for a comfortable lifestyle as well as, crucially, the ability to contribute to the family. In ensuring that they were economically active, and through their contributions, Hadi and Hashim demonstrated a wider attitude among young Muslim men in Luton, that working and providing for their family was an immutable symbol of 'good' masculine moral character.[10]

***

> When I was young, I knew I was different in this society. You knew you were living in England, that you were Pakistani, and that you were Muslim. You knew this because when you looked around […] there were people who weren't Pakistani or Muslim. But even though you know you're different, it didn't stop you finding out about the world.
>
> For example, I love football. I love playing football, I love watching football, I love talking about football. I always have. When I was young, Eric Cantona was my hero. I loved him more than I loved my family! The fact that he wasn't English, he wasn't Pakistani, he wasn't

Muslim, made absolutely no difference to me whatsoever. I wanted to be Eric Cantona as a child, and I still do!

– Amin, 23

Issues of race and ethnic difference heavily informed the formative lives of many of my friends in Luton. Moreover, ideas about themselves and the place that they occupied in society continued to manifest in attitudes towards wider society, and helped shape socialisation patterns. Eric Cantona – an iconic footballer known for his technical flair and charismatic personality – was a hugely popular figure in England throughout the 1990s. As Amin, a history teacher and poet, passionately pointed out, Cantona was a hero for many, regardless of one's religion, nationality or race. As a cultural figure, he was interpreted and reinterpreted in different ways by his devotees, including Amin. I asked him why he was in such awe of him:

He represented everything that I was and wanted to be. He wasn't [ethnically] English, neither am I. He was temperamental, so am I. He always got into trouble with the authorities, so did I. As a footballer, he was the best. Nobody came close. That's what I wanted to be. All my friends, whether they were Muslim, Black, White, Sikh or whatever, respected Cantona. Football brought us together. Cantona made us equal.

Amin had always dreamt of playing professional football. He watched the sport on television when growing up, played for numerous amateur and semi-professional teams, and even went as far as Manchester to watch matches. He was an avid reader of highbrow football magazines, and was generally very knowledgeable and passionate about the game. Amin was a goalkeeper, and first started playing while at school. Over the years, he played for a number of local teams and shone as a talent. Northampton Town FC, a nearby professional club, took him on after scouting him at a local match. Amin often mused about his 'failed' football career but, even though he hadn't 'made it' and decided to become a teacher, he remained hopeful. He stated that the two primary reasons why he failed were his temperament and his ethnicity. On the field, Amin had serious disciplinary problems. He could not remember the number of times that he had received a red card and been ordered straight off the pitch. Although his general demeanour was very pleasant and jolly, albeit quite manic, in his own words, once he gets onto the pitch 'something happens' to him. Amin had been sent off for excessive swearing, dissent against the referee, ungentlemanly conduct, dangerous play, violent conduct and many other eccentric charges (such as intimidating supporters). Off the field, he was full of shame for his actions. 'I just don't understand it', he admitted. 'I can't control my temper. I'm mad. Totally mad'.

The other reason he didn't make it as a professional goalkeeper was, according to him, because he was a 'Paki':

> How many Pakis play football? You can count them on one hand – ever! There's definitely a reason for it, beyond [South Asian] mothers not letting their little princes fuck up their GCSEs. You've seen how many Asians play football, they love it. Look at all these Asian leagues. You telling me that can't find good technical players there? Athletes? Bakhwas! [Rubbish!]

Bedfordshire amateur football and cricket leagues were full to the brim with Asian players. However, as Amin highlights, barely any of them became professionals in their respective sports. One notable exception was Monty Panesar, a cricketer of Sikh heritage from Bedford, who went on to star in the England team. Amin accepted Panesar's success but argued that it was an unfair comparison because Panesar was 'world class', a rare talent. He was always going to shine. In addition, cricket was a sport in which South Asians have contributed significantly throughout its history. The current teams from South Asia are some of the best in the world. Cricket, for Amin, was more tolerant towards Asians. Whereas football, he argued, was institutionally racist. It was the territory of dominant 'White Others'. When he was playing for Northampton Town, Amin played with mostly White and Black players. For his local teams, however, his teammates were almost exclusively Muslim of South Asian origin. Although he was always open to playing in multi-racial teams, and did so when he was at university, Amin had a preference to play with fellow Muslims:

> I've played for various football teams as a goalkeeper […] With the exception of Northampton Town, and my university team, all of them have comprised of mostly Muslims – usually local lads who love playing football. Although we played against non-Muslim teams, we were only comfortable with playing with other Muslims. A few tried to get into White teams, but they're always harder because you're Asian.

Another reason why we always played in [predominantly] Muslim teams, is because we all knew each other; it was always lads from the local area that you grew up with. Who are all mostly Muslim.

[...] Because I have a temper, I get into fights on the pitch. In those situations, you can only rely on your close friends to back you up.

Amin was quite obviously happiest when playing, watching or talking about football. He often travelled significant distances up and down the country (and sometimes abroad), in racially and culturally diverse arenas, to watch his favourite team. His living hero, Eric Cantona, was White and a non-Muslim. On the other hand, and quite paradoxically, his love and appreciation for football was ambivalently challenged by experiences of playing the game. He recognised, rather poignantly, that full participation in wider spaces of society was restricted for 'Asians', whom he perceived as excluded from the top echelons of the sport, despite possessing the necessary talent. Left somewhat disenfranchised, Amin settled for playing with his friends, locals with whom he grew up and shared a religion and ethnicity.

Participation in organised sport, particularly football and cricket, was an enormously popular leisure activity in Luton and its surrounding areas. Young South Asian men dedicated hours of their free time to organising sporting events, some travelling all over the country to attend amateur tournaments. Of course, these sports were not just popular with British Muslim men but are embraced in similar ways by different communities all across Britain. The participation in sport by young Muslim men in Luton was indicative of their wider engagement with communities and groups beyond their ethno-religious counterparts. It was also evidence that there was a will to integrate with the cultural mainstream, unlike their generational predecessors. However, as Amin's account illustrates, young Muslims often hesitated to consciously include themselves within that same cultural mainstream. Nonetheless, their persistence in organising matches, and following their favourite team, was an encouraging sign that, despite such perceptions, the desire to engage and interact with wider society remained undeterred.

A brief exploration of a group of friends tells a different but related story. The group consisted of four Muslim men of Pakistani origin, named Rashid, Faisal, Shakeel and Afzal, all of whom were in their early twenties. They grew up in the same neighbourhood and attended the same primary and secondary schools. One member of the group went on to university, whereas the other three pursued careers in retail. Shakeel and Faisal owned and ran a fast-food restaurant in nearby Aylesbury. Rashid worked locally for a large high-street mobile phone chain, while Afzal had been driving taxis since he graduated from university. I first met the group while waiting for a kebab at a famous grill house on Dunstable Road. After overhearing their conversation concerning holiday plans for Ibiza, I was intrigued and decided to introduce myself. The men received me warmly, seemed unperturbed by my interest and invited me to eat with them. Over the course of a series of conversations, the cultural complexities of their lives, and the varied and hybrid attitudes they possessed became more and more obvious to me. So many facets of their character and appearance were 'Western'; from the manner in which they dressed and the urban colloquialisms they employed in their speech to their obsessions with the latest electronic gadgets. At the same time, the 'East' was also palpable; from insistence on never missing Friday (*Juma'ah*) prayers in the mosque, to living in the same house as married siblings and their children as well as singing along to Hindi movie songs when driving in their cars. I asked them why they picked Ibiza as their holiday destination:

> Shakeel: We always go on holiday together. Since we were about eighteen or nineteen ... I can't remember now, but it's been a regular thing since then.
>
> Afzal: We went to Ibiza last year as well, and really liked it. So we decided to go again.
>
> Rashid [*pointing to Afzal and laughing*]: He really liked it because he was out of his face every night chasing butters [ugly] girls. Shakeel, on the other hand, was in bed by twelve!
>
> Shakeel [*to Rashid*]: There's only so much clubbing you can do. After a few days, you get sick of it!
>
> Shakeel: ... Anyway! As I was saying, we go on holidays a lot. It's easy because of cheap EasyJet tickets, and Luton Airport is just round the corner.

Despite insisting on teasing and scolding each other at every available opportunity, there was an obvious bond between these friends. It was ostensible in the way that they spoke to one another and the disguised esteem in which they held each other. The nature of the interchange between them was typical of any group of close friends. Moreover, the fact that they took advantage of cheap flights to visit European holiday destinations that specifically catered to young people, like Ibiza, Prague or Krakow, was demonstrative of their embeddedness within wider social and cultural norms in Britain. A venue such as Ibiza, which was notorious for hedonism – wild partying aided by the consumption of drugs and alcohol, and sexual activity – is seldom associated with impressions of Muslim youth in the West.

Once we had completed the meal, the oldest member of the group, Shakeel, insisted on paying for everyone, including me. When I protested, he retorted that as the oldest among his friends, he was obliged to pay. Especially when they were in the presence of a 'guest'. Further resistance forced Shakeel to explain to me how his act of generosity would 'look good in the eyes of Allah', and the divine blessings (*suaab*, Arabic *thawaab*) he would inevitably receive for footing the bill. Over the series of meetings and discussions that I conducted with this group, I learnt that their friendship occurred as a consequence of decades of close acquaintance between their parents. Their fathers all attended the same neighbourhood mosque over this period, and became friends in due course. Their families regularly visited each other's homes, and had even been on holiday to Pakistan together. They were the product of Bury Park's recent history and its distinct migrant community that had endured since the 1960s. What was most interesting to observe, however, was how this generation has modified old notions of how one lives and works in Britain as South Asians, to suit their more sustained exposure to wider British society and cultural practice. In this regard, they were constituents of an emerging generation managing to innovatively merge cultural influences from home and wider society, to create distinct and illuminating subjectivities particular to twenty-first-century 'multicultural' Britain.

Notes

1. See Hamid 2009, 2016; Meijer 2009; and Wiktorovicz 2006.
2. There is also a small but active Pakistani Shia community in Luton. However, my interactions with its members was relatively limited during fieldwork, due to lack of access.
3. See Robinson 2008; and Metcalf 1982.
4. See Hoque 2015.
5. Compare with Osella and Osella 2000.
6. Compare with Simon 2009.
7. See Deeb and Harb 2017; Werbner 1996; and Bayart 2007.
8. General Certificate of Secondary Education. This is the first formal qualification within the UK education system. Students are required to sit GCSE examinations when they are in their eleventh year of compulsory schooling.
9. Colloquialism for dealing illegal drugs.
10. See Osella and Osella 2000.

4
Religion

According to Olivier Roy (2004), young Muslims, particularly in diaspora, are being increasingly attracted to what he refers to as 'globalised Islam'. He argues that they are resisting the segregationist narratives of the modern nation-state. At the same time, they also resist the unfamiliar customs and expectations of their migrant families by overtly valorising religious identity and consciously associating with an international collective of co-religionists. Roy contends that the combination of globalisation, westernisation and the increase in worldwide Muslim diasporas has led to a reimagining of the international Muslim community (*ummah*). A global Muslim can mean either:

> Muslims who settled permanently in non-Muslim countries (mainly in the West), or Muslims who try to distance themselves from a given Muslim culture and to stress their belonging to a universal *ummah*, whether in a purely quietist way or through political action. (2004:ix)

Second and third generations in diaspora are particularly attracted to the doctrines of globalised Islam due to the fact that they are 'de-territorialised' from the Islamic heartlands. Moreover, they are attracted to this form of Islam due to the alienating cultural proclivities of both the host community and particular non-Islamic cultural preoccupations of preceding generations. Thus the appeal of being a member of the *ummah*, united by a common belief that transcends race, ethnicity and nationality is particularly powerful. In the post-colonial world, the '*ummah* no longer has anything to do with a territorial entity', but now must be 'thought of in *abstract* or *imaginary* terms' (2004:19 my emphasis). The egalitarian doctrines of Islam provide a comforting antidote to experiences of racism and social exclusion which are now part of everyday life in the West for

Muslims. Moreover, the notion that Islam is no longer fixed to a particular geo-political space adds substance to the doctrine of *ummah*, particularly among those who are living in diaspora communities.

Muslims in diaspora are also captivated by the notion of a 'glorious Islamic past' promoted by various reformist thinkers and groups – the idea that Islam and Muslims once spawned a great civilisation notable for its conquest of vast lands, its contribution to the arts and sciences, and responsible for the spreading of the religion to distant and remote places around the world. This pursuit for a 'pure Islam', devoid of polluting cultural influences, is consistent with the phenomenon of 'deculturalised Islam':

> The construction of a 'deculturalised' Islam is a means of experiencing a religious identity that is not linked to a given culture and can therefore fit with every culture, or, more precisely, could be defined beyond the very notion of culture [...] The new generation of educated, Western born-again Muslims do not want to be Pakistanis or Turks; they want to be Muslims first. (2004:22–5)

Although Roy's argument implies an effective rejection of established state narratives, it nevertheless does not suggest that young Muslims prescribing to a 'globalised Islam' will necessary evolve into terrorists. Rather, that young Muslims are re-appropriating their faith by challenging culturist approaches to the religion espoused by previous generations. At the same time, recourse to an abstract global community of Muslims reconciles perceived race and class disparities in their home nation.

***

Islam played a significant role in the lives of all of my informants. Everyone whom I met valued and prioritised their 'Muslim identity' above all else.[1] What this identity specifically meant varied from person to person but most were keen to show pride in their religion and solidarity with their co-religionists, both at home and abroad. This sense of belonging to an abstract creed (*din*) and imagined community (*ummah*) was one of the most prominent and powerful ontological forces shared by Muslims in Luton. Even so, there appeared to be a marked difference between the generations. Older men whom I spoke with agreed that Islam played a crucial role in their lives. Among the most avid attendees during the five daily congregational (*jam'aat*) prayers at the mosques, were men of retirement age or those approaching this age. Many such men claimed that they had worked hard all of their lives precisely for the privilege of being able to 'turn to God' in retirement and therein make amends for the mistakes accrued throughout a lifetime. It was also common for many older men and women to perform the Hajj pilgrimage to Mecca once they retired and were financially secure. Becoming 'religious' in old age, when family and work pressures relatively subside, was almost customary and expected of the older generation.[2] However, this trend is shifting. More and more young people in Luton were turning towards religion and a pious lifestyle. Moreover, young people were seemingly more attracted to reformist and revivalist variants of Islam than the traditional folk variants associated with South Asian Islam. These stressed the universalistic and 'de-cultured' expression of Islam, as Olivier Roy suggests, rather than the ethno-religious or localised variants that the older generation practised.[3] The majority of my British-born informants claimed that the 'village Islam' espoused by their parents or grandparents was 'backward' and inauthentic, and that they couldn't relate to it.

Luton has a varied and diverse Muslim community. The town's twenty-five mosques catered for a number of sectarian, ethnic and kinship-based congregations. Some are mostly frequented by those of Bangladeshi origin, others by Shia Muslims or the Salafi community and so on. Some of my young informants told me that they experienced pressure from home to only attend those mosques where members of their extended family and clan (biradari) made up a significant portion of the congregation, and/or where they controlled the mosque committee.[4] Even though most Muslims attended their preferred mosque most of the time, there were some for whom attending a mosque was a practical choice more than anything else. They might choose to attend the closest mosque for example or, notably, mosques where the imam spoke English. This latter point, and its concomitant implications, is crucial for our purposes in this chapter. Many young men with whom I spoke, harboured a mistrust for imams who were unable to speak English and could not explain the religion to them in a discernible manner. The Friday sermons (khutba), one of the distinguishing features of the widely-attended Friday prayer services (jum'ah), were given in either Bengali, Urdu or Arabic at most of the mosques in the town. The notable exception was the Salafi mosque. Some imams would on occasions embellish their speeches with some English commentary but, for the most part, English was a secondary language. Moreover, guest speakers would often be invited from South Asia or other parts of the Islamic world to give lectures and speeches, the vast majority of whom did not address the congregation in English. However, such events were, by and large, organised with the older generation in mind and consequently very few young Muslims would attend. In contrast, the Salafi mosque employed a policy of addressing the congregation primarily in English. On occasions where a guest speaker (many of whom came from Saudi Arabia) could not speak English, his speech was translated live by a supporting mosque official. Consequently, the Salafi mosque boasted a much higher percentage of young people from different ethnic backgrounds within its congregation than any of the other mosques that I attended.

In addition to the problems with communication at mosques, young people also complained about the 'magic' and 'superstition' associated with South Asian folk Islam. Many of my informants felt practices and rituals performed at these mosques, particularly those associated with the Barelvi Sufi sect, were cultish and cultural, rather than anything associated with the 'pure' teachings of the Qur'an or the practices of the Prophet Mohammed. Barelvi Sufis are particularly devoted to the Prophet Mohammed, arguing that he is still alive and present (*hazil/nazil*) in the world. In most Berelvi mosques, a vacant chair is placed at the front of the prayer hall during prayer times to indicate that the Prophet is also praying with the congregation. This doctrine was fiercely defended by the older generation and other committed Barelvi Sufis, especially in light of the persistent attacks from revivalist sects like that of the the Salafis. To almost all of my young informants, this ritual was irrational and based on superstition rather than reference to proper Islamic scripture. They claimed that such practices were for uneducated, rural Muslims in South Asia. In a sense, debates around the clash between sober 'scholarly' Islam and syncretic 'folk' Islam are not new, and anthropologists have been keen to highlight this apparent theological tension in the Muslim world for quite some time.[5] More recently, others have argued that Sufi orders[6] have resisted and survived the disenchantment of modernity and adapted their practises to suit the times[7] or witnessed a growth in their congregations as a consequence of it.[8] However, whatever the anthropological literature suggests, in Luton there was an apparent chasm between those who were born there and those who were not. More and more young Muslims were rejecting the folk Islam of their parents and grandparents, and relating more to the literate, puritanical Islam of the reformists. They were not doing this without encountering problems. Similar to the insights garnered by Samuli Schielke among young men in northern Egypt, Luton's young Muslims likewise struggled to live pious lives 'according to the book'.[9] Nevertheless, unlike Schielke's informants, Islam also formed the basis of nascent *political identities* for young Muslims in Luton that I suggest are symptomatic of an emerging generation of Muslims living in the contemporary West, and native to it.

***

Even though revival or reform Islam appeared to be most attractive to my young informants, the institutional presence and influence of traditional South Asian Sufism remains strong for the time being. As I have mentioned, the majority of the mosques remained firmly in the hands of the 'elders' (exclusively men). This meant that community elders recruited imams, ran the associated seminaries (*madrasas*), were responsible for estates and maintenance, and organised religious and cultural events. Thus all religious learning and services very much remained in the control of the older generation (apart from the Salafi Mission, which I shall return to later).

Growing up, all of my informants attended the mosque for congregational prayers, religious festivals and seminary school – where they were taught how to perform the daily prayers (*salat*) and other important rituals, in addition to learning how to read the Qur'an in Arabic. Young boys and girls of schooling age were regularly packed off to the mosque after school and on Saturday mornings. This was a practice taken directly from many parts of the Islamic world, not least South Asia.[10] For many, the experience of attending lessons at the mosque made an indelible impression. It was seldom positive. They complained of being regularly beaten by the imam if their work (*sabbak*) was unsatisfactory, if they misbehaved or even when the imam 'just felt like it'.

> The imam used to make us memorise the Qur'an and practice *namaz* [daily prayers]. He couldn't speak English, and only spoke to us in Urdu, which a lot of us didn't understand. Also, he never explained to us what any of the Arabic meant. We were just expected to memorise it all. For a long time I thought that was it, that was all Islam was – praying and memorising the Qur'an [...] Actually, it put me off Islam [for a bit] because when we didn't get our *sabbak* [daily tasks] right, we were beaten and told we would go to hell if we kept getting it wrong. Now though, I've found out what Islam is really about by speaking to older cousins and people at the mosque, and reading books about Islam in English.

> – Abbas, 18, A-level student

Abbas's account of his experiences of *madrasa* were typical. Corporal punishment was a common pedagogic practice that most parents approved of. Moreover, chapters of the Qur'an were learnt by rote in Arabic, without translations or commentary on complex concepts and verses. Imams, more often than not brought over from South Asia for low salaries, could not speak proficient English and thus could not relate to their students. Despite this, however, and quite surprisingly, Islam remained important to many of my informants' lives in adulthood. Abbas, who claimed he was 'put off' Islam, argued that his experiences were just the product of being taught in a Pakistani way and that future generation should be taught differently:

> In Pakistan, this is how they learn about Islam [...] They know Urdu, so they can ask questions and get a better understanding. But here, most of us weren't interested in going to the mosque after school, and when we [were there] our questions never got answered because the iman didn't care or we couldn't understand him [...] We should get taught about Islam by people who can speak English, so we can understand our religion better.

Abbas didn't harbour any lasting antagonism towards Islam. In fact, he was making concerted efforts to indeed try to understand the religion better. Like so many other young men in the town, he was attracted to the Salafi Mission and, in particular, to the preaching of Khidr. He was a regular attendee at the mosque for Friday prayers and also came to events and lectures organised by the mosque. He especially enjoyed Khidr's lectures and sermons because, he claimed, Khidr was from Luton and spoke perfect English as well as perfect Arabic. Abbas also felt Khidr wasn't a hypocrite, but a kind, sincere and pious man, making huge sacrifices for the sake of his religion. Despite this, Abbas was not quite sure whether he himself was a Salafi. 'It's a big thing, a lot of hard work', he told me. 'I'm not sure I can be like that, but hopefully one day'. Abbas was concerned that the austere discipline and abstinence required to fully 'practise' Salafism was beyond him for the time being. He also claimed that he didn't quite possess 'knowledge' about Islam, and that he was still re-learning all the basics. Coming to the Salafi mosque, therefore, was a means to re-educate himself about Islam. Being 'pious' and one day like Khidr clearly appealed to him, however. More so, it seemed, than emulating the religious teachers of his childhood.

***

Zulfiqar was an IT consultant in his mid-twenties, who often visited his friends at the MSP offices. He wore a very large black beard, along with a long white Arabian-style thobe to work (even in winter). Zulfiqar also owned a selection of the attars (perfumed oils) produced in Saudi Arabia. Whenever Zulfiqar was in the office, a pungent waft of his chosen scent for the day circulated the building. On occasions, this was oppressive. He claimed that wearing attar was 'sunna', something that the Prophet did and encouraged. Salafi Muslims are particularly keen on Prophetic mimesis. For them, like most Muslims, Mohammed was the living guide on how to live one's life according to the 'Book'. However, Salafis adopt a literal meaning to this, arguing that everything that the Prophet did should be emulated as he did it, regardless of the rational, logical or practical problems that this may cause. Zulfiqar's choice of fragrance, therefore, could be seen as an act of piety and religious obligation (rather regrettably for the rest of us). But this mimesis as an act of piety was a certainty from *his* point of view. When I asked him if he had ever ridden to work on a camel, as the Prophet did, he replied: 'The scholars have said that the car is halal, Ashraf'. Then, after a slight pause, he chuckled: 'maybe we can order one for lunch?' Zulfiqar was a devout Salafi Muslim. He prayed five times a day, almost all of which he performed at the mosque during the congregational hours. He was a regular at the dawn (*fajr*) prayers, which in the summer months meant waking up around 3am. During Ramadan, he would fast the entire month in addition to his daily prayer routine. Then, when Ramadan finished, he would continue holding the voluntary (*nawafal*) fasts after Eid because 'the Prophet recommended it'. It is also said to be Prophetic tradition to fast on Mondays and Thursdays, precedents that Zulfiqar often also observed. He had been to Mecca on the Hajj and the minor (*Umrah*) pilgrimages several times, the first as a child. He paid his obligatory religious alms (*zakat*) promptly every year, he never took nor received interest in any form nor did he socialise with any unrelated women (*mahram*, i.e. socialising only with wife, mother, sister, daughter, niece). He was also friendly, well-meaning and light-hearted.

Zulfiqar had not always been so piety minded, however. Like Abbas, he was disenfranchised by the 'village Islam' of his parents. He was yet another Muslim of his generation that was highly critical of the *madrasa* religious schooling system, arguing that it needed serious reform. Zulfiqar claimed that he had not learnt anything of note at madrasa, beyond the act of reading Arabic and memorising chapters of the Qur'an. He argued that there was no intellectual development, and that important and classical Islamic subjects such as language (*lugha*); history (*tarikh*); law (*kanun*); jurisprudence (*usool-ul-fiqh*); creed (*aqidah*); theology (*kalam*) and methodology (*manhaj*) were not taught to students at the madrasa. For Zulfiqar, this lack stunted the progress of students and their knowledge of Islam. Furthermore, he noted that his teachers could barely speak English and couldn't adequately explain the complex meanings of the scripture. He didn't know it at the time but, looking back, he felt that he was completely misguided by his religious teachers growing up. His attitude towards the religious learning he gleaned from his family did not fare much better. 'They just wanted me to pray, and fast, and stay away from girls', he declared. 'They didn't really care about gaining knowledge of the *din* [religion]. It was all about being a good, *sharif* [respectable] boy'. Zulfiqar's parents maintained close ties with Pakistan and loosely followed the Berelvi sect. Despite this, they were not zealous, and encouraged their children to be good, God-fearing Muslims above any sectarian loyalties or allegiances. Consequently, Zulfiqar's conversion to Salafism in his adulthood (or 'reversion', according to him), did not cause tensions in his relationship with them. This is because the Salafi lifestyle is based on personal piety and ritual discipline which, for his parents, were the hallmarks of a 'respectable boy'. His parents were thus not concerned with the theological underpinnings of his new beliefs. All they cared about, according to Zulfiqar, was that he was ostensibly praying and fasting, and not chasing girls.

Zulfiqar converted to Salafism while he was at university but said he was aware of the 'brothers' (read: congregation) in Luton while he was growing up. At the time, he just thought Salafis were like any other Muslims but just more committed to religion. It was only when he engaged in deeper conversations with practitioners that his opinions and lifestyle began to change. Prior to going to university, in his words, he lived an 'ignorant' (*jahil*) life. This term forms a regular feature of Salafi discourse, made in allusion to the 'Age of Ignorance' (*jahiliyya*) of pagan pre-Islamic Arabia, which directly preceded the era of the Prophet Mohammed that was marked by the revelation of the Qur'an. Despite smoking cannabis on rare occasions with his friends, and talking to girls at school and college, Abbas claims he did not do much wrong, as he always tried to pray and fasted without fail during Ramadan. It was only when he gained knowledge (*'ilm*) of Islam that he was compelled to change his outlook and practices:

> When I first started university, I only knew what the local imam and what my parents had taught me, [but] I was very interested in religion. I was questioning a lot of things as I was growing up. A lot of the things the community did didn't make any sense to me – the *pirs*

[saints] and the magic and all that – I just didn't believe in all that. It just seemed really backward and stupid [...] Some of my friends [at university] were meeting up with these brothers from the [student union] Islamic Society who would take them to Islamic talks. They asked me to go, but I always refused until one Ramadan when, for some reason, I decided to go. It was a talk on Islam and economics, and I remember being really confused at the title as those two things don't mix, at least I thought they didn't [...] Looking back now, that decision completely changed my life.

Going to university was a life-changing experience for Zulfiqar. He became involved with Salafi missionary activity on campus (*da'wa*), and started living a more puritanical life, giving up all his previous excesses. He made lasting relationships with devout Salafis from all over the country, particularly Birmingham, where the sect has its largest presence in terms of numbers and activity. Zulfiqar often made regular visits to Birmingham to attend lectures and seminars, and catch up with friends. After graduating, he returned to Luton and continued living with his parents. Once back, he decided that he would also engage in the '*da'wa*' on the local scene, transferring the missionary work he had embarked upon on campus to his home town. This was highly important to Zulfiqar because, as mentioned previously, the act of *da'wa* is a religious obligation for Salafis. As per their belief in mimesis, this importance stems from the traditions that hold the Prophet Mohammed spent his lifetime proselytising. Luton was also fertile ground for this kind of Salafi *da'wa* activity. The town boasts a large Muslim population, many of whom are British-born, English-speaking and seeking alternatives to the impenetrable perspectives on Islam offered by more mainstream institutions. Zulfiqar thus teamed up with the other members of the Salafi mission (some of whom he knew through community networks prior to his conversion), and set about his work. Despite his obvious zeal, clear intentions and experience on the national *da'wa* circuit, Zulfiqar was not very good at it. He found it difficult to bond with youths that frequented the offices, who couldn't relate to his 'geeky' mannerisms and overly complicated and dry theological arguments. By his own admission he was also quite lazy. He found it challenging to motivate himself for preaching, often comparing himself with the likes of Khidr and Ammar, and concluding that he was just simply not on their level. For the most part, however, Zulfiqar was happiest going to the mosque, learning about his religion, and playing video games in his spare time. It was apparent that he enjoyed the 'black-and-white' coherence and certainty that the Salafi doctrine brought to his life – trying to emulate the spiritual example of the Prophet in all that he did even though, at times, this caused him strife and self-doubt. Following the Prophetic tradition was the 'safest' option, because only the Prophet knew the true meaning of the Qur'an and, by implication, Islam itself. Importantly, he was a member of a quietist branch of the Salafi sect who maintained a conscious distance from political affairs and thought Luton, in his own words, was 'the best place on Earth'.

***

Although my friends at the MSP were very active and well known in the town, and generally enjoyed a good reputation for their piety, manners, youth and rehabilitation work, they were not the only Islamic revivalist group operating in Luton. Also significant was the presence of Tablighi Jammat (TJ) another transnational revivalist organisation.[11] TJ, like the Salafi sect, also maintain a quietist stance and distances itself from politics. In terms of religious or missionary activity, TJ is again similar in its theological and methodological approach to the Salafis. There is a strong emphasis on personal piety and daily ritual observance, and an obligation to carry our *da'wa* wherever its members are, regardless of their social or economic circumstances. Moreover, TJ Muslims that I spoke with believed that Muslims in general had lost their way in the modern world. This was as a result of the cultural effects of Westernisation and colonial domination, which led to Muslims abandoning the true essence of their faith. This, they argued, resulted in the wrath of God that was undeniably demonstrable by the relatively lowly plight of the contemporary global Muslim community (*ummah*). By way of an antidote, TJ Muslims advocate a 'back to basics' doctrine, where 'non-practising' and 'lapsed' Muslims were gently encouraged to attend the mosque, seek penance and regularly perform religious rituals and obligations. By reverting to a basic code of personal ethics and discipline, thereby creating a more pious entitled community in the eyes of God, Muslims could build a foundation for eventual revival and a return to past glories.

Ostensibly, there is seemingly not much difference between TJ and Salafi Muslims. Both emphasise and value personal piety, ritual adherence and a yearning for the 'purity' of the past. Similarly, many TJ Muslims that I came across were invited to the movement by committed preachers (*da'is*), and many of the 'converts' were hitherto lapsed in their faith or had been disillusioned by traditional alternatives. However, there were distinct and insurmountable differences between the two sects in the realm of creed (*aqidah*) and theological methodology (*manhaj*).

Much of this discussion is beyond the scope of this book but, for our purposes here, it is worth noting that Salafis claim that their creed is the only one that is totally devoid of human innovation (*bid'a'*) and polytheistic practices (*shirk*). They argue that the classical scholars of Islam inserted their personal reasoning and biases into sacred rulings. This was unacceptable according to the Salafis, since Islamic scriptures were divine and could not be tampered with by human agents. Consequently, Salafis interpret the scriptures literally, rather than the analogous and hermeneutic approach taken by every other Sunni and Shia legal and theological school throughout the ages. Since TJ Muslims, by and large, follow mainstream Sunni schools of law (mostly Hanafi), Salafis reject them as 'deviant'. Salafis also argue that scholars rulings on issues where there is no direct evidence from divine sources is akin to polytheism, both on the part of the misguided scholar and also on the part of the uncritical devotee, who has rendered trust in matters of religion to a fallible human agent. The logical charge for Salafis, therefore, is that such individuals were following themselves and/or other humans, fraught with imperfections, rather than the infallible example of the Prophet Mohammed. This, they argue, is associating partners with God and wholly against the Prophetic message of monotheism (*tawhid*). Differences aside, my friends at the MSP and TJ were both politically quietist. However, as we shall see, not all Salafi groups in Luton were politically disengaged and neither were all political movements there secular.

Luton was also a centre of activity for two of the most notorious 'Islamist' movements that were openly recruiting in the town at the time. One of these organisations, formerly known as Al-Muhajiroun (AM), has since been banned in the UK after an amendment to the Terrorism Act in 2010. The other, Hizb-ut-Tahrir (HT), holds a similar political vision but its activities remain legal in the UK, for the time being at least. Both AM and HT call for the re-establishment of the Caliphate (*Khilafah*) – an imagined Pan-Islamic nation-state where sharia law forms the basis of the constitution and public life.[12] Activists from both groups argue that only God's law should be applied on Earth, as man-made laws are prone to error and, of course, following them is akin to polytheism (*shirk*) because rule (*hukma*) is the exclusive domain of God. The leader of AM, Omar Bakri Mohammed (OBM), was once the leader of HT in the UK. AM was created in the mid-1990s after a disagreement on HT's methodology (*manhaj*) among its then leadership. OBM took a literalist view and argued that the establishment of the Caliphate was a religious obligation anywhere in the world, including Britain. HT's position on the other hand had always been to establish the state in a Muslim-majority country for practical and strategic reasons. Both organisations cited evidence from the Qur'an and Hadith literature to justify their political mission and claimed that working for the re-establishment of the Islamic state is the foremost religious obligation (*fard*) for all Muslims living anywhere in the world. The absolute necessity for living under an Islamic state was further predicated on the argument that living a righteous and pious Muslim life was only possible within such a polity. Both argued that it was incumbent on all Muslims to work for this goal above all others and that, through this, they would bring an end to the humiliation of Muslims around the world that began with the advent of colonial domination of Islamic lands and its people. This particular expression of anti-colonialism, from my observations, maintained great traction with many of my informants but, curiously, it compelled very few of them to join either organisation. In fact, both HT and AM were widely ridiculed by the majority of my informants who knew of them and what they stood for. Many dismissed their agenda as idealistic and impractical, while others questioned the theological legitimacy of such a position. Consequently, both these groups wielded a very marginal influence among the youth or the Muslim community as a whole.

***

> People say America is the Great Satan. They're wrong. Britain is the Greatest Satan. Look what they've done to the world. The Americans are amateurs compared to the British [...] They [the British] went around the world and took whatever they wanted. Thieves and murderers. Muslims should take back what is theirs [...] Look at what they're doing to Muslims today – [Muslim] sisters being attacked in the street because they wear the *hijab* [headscarf], [Muslim] brothers can't get a job unless they shave their beards [...] You're Muslim aren't you? Do you think they [non-Muslims] like you here, bruv? [...] The *kuffar* [non-believers] are at war with Muslims because they want our wealth and stop the spread of Islam [...] Allah says in the Qur'an: 'They may plot and plan, but I am the best of plotters and the best of planners'.

> – Hamed, 25

There were very few members of HT and AM who were active in Luton at the time of my research. Of those members who were, a significant portion came in from other parts of England to preach. Bury Park's high concentration of Muslims made it a fertile ground for recruitment, at least in theory. HT activity in the town at the time was, by and large, mostly out of sight. They chose to recruit members through social networks and staged events. While AM also employed a similar strategy they ran, in addition, a 'da'wa' stall on Dunstable Road in Bury Park. Almost every day, from late morning to early afternoon, a group of bearded, predominantly South Asian men, dressed in robes, combat gear or a combination of both, stood around this stall looking to strike up a conversation with passers-by. They would disperse into smaller groups of twos or threes at busy periods. Sometimes, if the debate was particularly contentious, the various clusters would unify with all focusing on the battle between one set of opposing individuals. Such debates have been known to get rather heated on occasion. Most of the time, however, people passed by without noticing them. On the stall itself, one could find a rich assortment of fundamentalist literature. Some incited violence, others were anti-semitic, and a few sought to provide critiques of capitalism. At the same time, one could also find leaflets and pamphlets promoting ritual piety, the importance of monotheism and Qur'anic exegesis. By far the most salient topic, however, was the Caliphate. More specifically, on how it was a fundamental duty of all Muslims to re-establish the Caliphate wherever they find themselves in the world (as contrary to the HT methodology of establishing the Caliphate in the Islamic world for practical reasons).

Activists at the stall spent most of their time talking among themselves. It appeared that they shared a close bond with each other. Some of them were close friends and were constantly eager to remind me that they loved each other (as brothers), 'for the sake of Allah'. Others were less interested in the social aspect of the work but committed solely to the 'da'wa'. In my first encounter with members of AM, they were not aware that I was a researcher. I approached the stall out of curiosity and began to scan through the literature, while the others were either talking to passers-by or to each other. After a while, I was approached by an individual who introduced himself as Hamed. He asked me if I was interested in 'the Truth'. We chatted for a while about the need for the Caliphate, the West's war on Islam and the urgent need to join a political party. Anticipating a longer sermon, I disclosed to him that I was actually a PhD student, studying Muslims in Luton, and would very much like to talk to him about his beliefs. I suggested that we meet for lunch after his duties were completed, and he agreed.

We arranged to meet at a fried chicken shop a few metres away. On arrival, I was immediately greeted by over a dozen other AM activists, some of whom were present at the stall earlier. After some pleasant exchanges with the activists, I was ushered to a table with a few other activists, and immediately offered food. Hamed was of Bangladeshi origin and had lived in Bury Park all his life. He studied Computer Science at a university in London and, upon returning to his home town, starting questioning the role and relevance of Islam in his life. After a period of investigation – attending lectures at the mosque, reading Islamic books and contacting prominent local religious personalities – he decided to join AM. I asked him what ultimately swayed his mind:

> The Muslim *ummah* is divided. Our Muslim brothers and sisters are being oppressed every day. As a Muslim, it's my duty to defend the honour of the *ummah* [...] We have a glorious history; we conquered so much of the world and led the world in so many fields, we were able to achieve all of this because of Islam [...] We have lost that power now, and every Muslim should work in the path of Allah to get that back. I work with these brothers to re-establish the Islamic state because Allah orders me to do so.

Although this response seemed somewhat rehearsed, it struck me how similar the recruitment methods and rhetoric employed by Hamed and his comrades were to those employed by HT. As mentioned earlier, the key difference between the two groups was on methodology (*manhaj*) and creed (*aqidah*). HT took a 'rational' approach to all of their activities, even in creedal matters. AM, on the other hand, argued that human rationale is flawed and akin to grave innovation (*bid'a*). When fulfilling one's religious obligations, they argued, one should adhere strictly to the text without besmirching its purity with human inference. In this regard, AM members self-identified as Salafis or, more precisely, Salafi-jihadis. This attribution was, however, fervently contested by my informants at the MSP – who themselves were members of the quietist and most popular branch of global Salafism. My colleagues at the MSP argued that AM members were 'misguided' and 'lacked knowledge about the *din* [religion]'. They took it upon themselves to engage in debates with the AM in order to bring them back into the fold of 'orthodoxy'. Consequently, delegations from the Salafi mission were frequently sent to the stall for the purpose of debate. Without fail, however, these 'debates' digressed into highly charged inter-sectarian public slanging matches, with either side refusing to yield. Despite the evident lack of progress, the debates continued. I always wondered to what extent the two groups were actually trying to persuade each other, rather than merely parading their self-perceived superior knowledge and consequent piety for the public gaze.

Back at the chicken shop, we moved on to discussing the difference between HT and AM. I suggested to Hamed that, to me, his mission seemed identical to that proposed by HT. The only difference being that AM members were quite content to establish the Caliphate in the UK, for example, whereas HT members were not. Hamed replied with a theological justification discussed above but also cited the credentials of his leader, OBM. He argued that OBM was once the leader of HT in the UK and that it was always his intention to establish the Caliphate in the UK. Moreover, Hamed added that this was the reason why HT's global leadership (based in Jordan) decided to expel him from the role. In the 1980s and 1990s, OBM was responsible for the recruitment of hundreds of young British Muslims to HT. By 1996, his charismatic hold on the group in the UK began to cause consternation within the global leadership of the party. They feared that a cult of personality, centred around OBM, was replacing loyalty to the organisation. In addition, OBM's rogue, Salafist-inspired methodology to re-establish the Caliphate went directly against the leadership's strategy of focusing efforts in the Muslim-majority Middle East – an area they referred to as the 'majaal' (domain). For HT, its role in the UK was simply to garner support for the ideology of the party and to recruit high-profile members of the international Muslim elite that frequented its shores. The idea was to send these individuals back to their countries of origin (in the majaal) where their prominence, social capital and further recruitment activities could contribute to the eventual goal of staging an 'Islamic revolution' through a military coup.

Initially, AM was a non-violent organisation, imitating HT's approach of calling Muslims into action (da'wa) and encouraging radical political engagement. However, after 9/11, the organisation underwent a methodological reformation. In the aftermath of 9/11, OBM declared himself a Salafi and defended the attacks as a necessary military technique in the obligatory jihad against the West. This positon only intensified leading up to the 7/7 bombings in London in 2005. Following the bombings, OBM left the UK for Lebanon. While there, the Home Office identified him as a key agent in the radicalisation of British Muslims (such as the 7/7 bombers), and informed him that he could no longer return to the UK. At the time of research, OBM had been absent from the da'wa circuit in the UK for three or four years, but his influence remained as potent as ever. 'He is our sheikh. Our leader and our teacher', insisted Hamed. 'He instructs us in Islam using the Qur'an and Sunnah as the only source, and his knowledge of the din [religion]'. It was obvious that Hamed was impressed by his 'sheikh'. He frequently referred to OBM's wide training (in Saudi Arabia), moral character and wisdom on all matters.

At the time of writing, OBM is serving a six-year sentence in a Lebanese prison for terrorism charges and links to Syria's Al-Qaeda-inspired Nusra Front. Since his exile from the UK, AM has taken on many guises under the leadership of Anjem Choudary – a British-born Muslim of Pakistani heritage and loyal lieutenant to OBM. At the time of research, AM operated legally in the UK. In 2010, however, the organisation was proscribed under the Terrorism Act 2000. Despite this, the organisation has continued to operate under various aliases, promoting violence against non-believers (*kuffar*), and encouraging Muslims to engage in jihad against the forces of the West. In 2016, Choudary was also imprisoned for supporting and inviting others to support and/or join the Islamic State of Iraq and the Levant (ISIL/ISIS), after the latter declared itself to be a *bona fide* Caliphate in 2014. Since then, a number of British Muslims who were members or affiliated with AM have travelled to Iraq and Syria to join ISIS. I never saw Hamed again after our meeting, and I often wonder what may have happened to him, fearing the worst. His austere views and chilling anger towards 'the West' left me quite shaken. I found that I simply couldn't reason with him. Although his interpretations, and those of his friends, were very marginal in Luton, the potent sense of conviction, coupled with a highly politicised explanation for the lowly status of Muslims in the contemporary world, clearly chimed with some, both here and abroad.

***

HT and AM are marginal voices within the Muslim community, not just in Luton but among British Muslims in general.[13] Although their ideas are radical, they are predicated on the fundamental belief that the holy scriptures of Islam (the Qur'an and Prophetic Traditions/Hadith) are the only source of authority for Muslims. This implies the rejection of any non-Islamic cultural influences within the sphere of belief and conduct. Although AM and HT adopted an extreme interpretation of this doctrine, all Islamic revivalist and reformist movements (including TJ and quietist Salafis) seemingly share the same concern vis-à-vis the need to 'purify' the religion, and restore it to its previous glories. These ideas profoundly resonated with almost all of my British-born informants in Luton. The abstraction of Islam into a de-culturised, ethical and political identity was highly appealing to young Muslims seeking alternatives to the ethno-religious identities held by their parents. In addition, it also appears that, for a disturbing proportion of my young informants, wider perceptions of Muslims in the era of the 'war on terror', and the constant barrage of negative press relating to Muslims,[14] had consolidated a sense of 'Otherness' in relation to wider British society. My informants constantly claimed to be 'Muslim' above and beyond all other identity markers, and emphasised a will to demonstrate to non-Muslim others that Islam was not a religion of terror or violence. Many argued that Islam was a peaceful 'way of life' that could be applied anywhere in the world – whether one was in the minority or majority – and that this was Islam's 'beauty'. Moreover, and quite paradoxically, this 'beauty' required the stripping of idiosyncratic cultural syncretisms borrowed from South Asia in order to be realised.

The appeal of revivalist Islam, therefore, is two-fold: firstly, in the sphere of ritual piety and ethical practice; and secondly, through conceptions of group solidarity that transcend the nation-state. Young Muslims welcomed the existential security of a 'back to basics' doctrine of certainty, because they were necessarily disenchanted after centuries of epistemological deviances. They warmed to the powerful idea that Muslims should do 'as the Prophet did', and reject all of the innovations that came after him. Furthermore, the egalitarianism and inclusivity implicit within the concept of '*ummah*' played a crucial role in identity formations. For many, concern for the plight of the international community of believers was not only a religious obligation but a political reality. The 'war on terror' and its consequent geo-political effects were palpable in the conversations that I had with so many of my informants. Events in Palestine, Iraq, Afghanistan, Kashmir and so on, were important because co-religionists in these places were perceived to be battling against the forces of oppression, poverty and violence. My informants identified deeply with these strangers in distant, unfamiliar lands, and wanted something to be done about it. Some of them, like Hamed, wanted to fight neo-imperialism through establishing a bellicose Islamic state, others wanted to raise funds for charitable causes in the Muslim world, while others still concluded that prayers (*salawat*) and constant mindfulness (*fikr*) were the best options during such dark political times.

As I have argued in this book, such events and developments have seemingly given rise to novel forms of religious identity intrinsic to Luton or, by extension perhaps, to Britain. In accordance with Talal Asad's famous intervention, we may be witnessing the rise of a localised 'discursive tradition' (1986), where British Muslims are striving for a sense of text-based orthodoxy that suits their particular temporal and genealogical positionalities as a post-colonial diaspora. However, as Samuli Schielke (2009) rightly notes, not all Muslims are pious all of the time and, equally, not all transgressors are morally apathetic. Rather, Muslims everywhere are in a constant process of negotiating conflicting and incoherent 'moral registers' in the practice of everyday life, none more so than those living in societies where Islamic values are marginal or 'Other', as is the case with Britain. Although 'the text' was undoubtedly important for all of my informants, many of them (in fact the vast majority) did not consciously adhere to religious rulings in everything that they did. One such inform- ant, Joynul, was a seventeen-year-old young man of Bangladeshi origin, whom I met at a Bangladeshi-majority youth centre in the heart of Bury Park. In his own words, Joynul didn't 'practise' Islam but held aspirations to do so in the future. Although he frequently visited Bangladesh with his family, and shared a close relationship with many relatives there, he did not identify as a Bangladeshi in the same way as his parents did. Curiously, he was also uninterested in claiming to be English either, arguing that he couldn't be fully English as he was not White, but a Muslim. Rather, he was most comfortable to be referred to as a 'British Muslim'.

> Do I think I'm British? Yeah, I've got a British passport, so I'm British, but I'm not English. I'm a Muslim [...] I don't pray five times a day or anything, [...] I'd like to when I get older, but I'm still Muslim, and I'm proud of that, even if people look down on us in England.

Clearly, Joynul's ethnic identity did not matter too much to him, as he did not fully identify as either a Bengali or an Englishman. Instead, he priori- tised his religious identity without being, as he claimed, 'a religious person'. Thus, Joynul's Muslim identity was not just about being religious, but it suc- ceeded in creating a cognitive space where his ethnic, religious, sociologi- cal and political differences with wider society could be reconciled. Joynul did not feel at home with his peers in Bangladesh. At the same time, while he conceded that he may be 'more Westernised', he nonetheless refused to be labelled 'English'. In this regard, as we have seen, his views correlated with the majority of my British-born informants. I asked Joynul whether he thought his Bangladeshi cultural roots complicated a strictly Muslim iden- tity based solely on adherence to the Qur'an and Prophetic Tradition:

> Not really. Every one there [Bangladesh] is Muslim, and all the Bengalis in Luton are Muslim as well. But I don't think I'm Bangladeshi, just like I'm not English. Bangladeshis of my age are different to the ones in Bangladesh, we're more westernised [...] But you can live here and still be Muslim, can't you? You can be Muslim anywhere. It doesn't matter where you're from. That's why

America is trying to fight Muslims everywhere. They [America] know how big the *ummah* is.

His final remark was also telling. Joynul argued that Islam can be observed anywhere and this for him was the reason why America had declared the 'war on terror'. Even though the universal time-space applicability of Islam was a common trope, to jump from religious observance to geo-political commentary was curious. Nevertheless, this was certainly not unusual among Joynul's peers in Luton. Young Muslims often conflated a sense of being Muslim with affinity to the political affairs of fellow Muslims around the world. The more pious ones, such as Zulfiqar, justified this 'concern' (*fikr*) for fellow Muslims by citing scripture, such as the following hadith: 'The *ummah* is like a body; if one part of it is wounded, the whole body responds with sleeplessness and fever'. In contrast to Zulfiqar, however, Joynul was not a member of any revivalist or reformist group. Nonetheless, he still advocated the need for solidarity with fellow Muslims, wherever they were. Moreover, even though he was not a 'practising' Muslim, he not only self-identified as a Muslim but claimed it was his *primary* identity – above and beyond ethno-linguistic or nationalist allegiances of his parent's generation. In this regard, his views were consistent with the vast majority of my British-born informants who were mostly comfortable with identifying as British, but were quick to point out that their Muslim identity was just as important to them, if not more so.

Notes

1. Compare with Considine 2018:7; and Gest 2010:104–8.
2. See Gardner 2002; Watson 1977; Ballard 1994; Jacobson 1998; and Inge 2016.
3. See McLoughlin 2013; and Gest 2010:104.
4. See Werbner 2002.
5. Compare with Gellner 1981; Robinson 1983; Das 1984; Van der Veer 1992; and Osella and Osella 2013.
6. Sufism is the mystical dimension of Islam that prioritises personal 'union' (*wahda*) with God through ritual meditation (*dhikr*) and, often, obedience to a spiritual master (*shaykh*), although not always. Sufis believe in the 'unseen' (*al-ghayb*) and intercession of both living and dead saints (*awliah*). These ideas and practices are regarded as heretical (*bid'a*) by revivalist and reformist sects.
7. See Gilsenan 1973.
8. See Werbner and Basu 1998; and Chih 2007.
9. See Schielke, 2009:S32.
10. See Bledsoe and Robey 1993; Sperling 1993; Sanneh 1997:117–46; Brenner 2001:39–84; Zaman 1999; and Malik 2006.
11. Compare with Metcalf 1982; Sikand 2002; Horstman 2007; Janson 2014; and Pieri 2015.
12. See Taji-Farouki 1996; Mandaville 2001; and Wiktorowicz 2005.
13. See Hamid 2016.
14. See Morey and Yaqin 2011.

Conclusion

In her seminal paper 'Zones of Theory in the Anthropology of the Arab World', Lila Abu-Lughod concludes that the focus on the Muslim world should move beyond the concerns of 'classical' anthropology, and incorporate 'transnational cultural forms, global communications, labour migration, and international debt' (1989:300). She identifies three 'theoretical metonyms' in the anthropological study of the Arab world at the time: segmentary lineage theory, harem theory and Islam. Although some of this historical focus can be attributed to the wider development of anthropology as a discipline in the Middle Eastern context, this is also in keeping with, she argues, Orientalist scholarship and the particular power relations invoked by the mores of colonialism. Thirty years on, the anthropology of Islam has taken some heed for her vision. Studies of Muslims, in transnational/global contexts where political and economic 'push' and 'pull' factors have given rise to new social forms beyond the Islamic heartlands, have been burgeoning.[1] This book is a contribution to this emerging literature.

Recent anthropological debates on the anthropology of Islam have focused on a 'lived' approach when studying Muslims.[2] This challenges Talal Asad's view that Muslims are defined primarily by their religion, which unites them discursively through time and space (1986). In the former view, Islam forms merely an important aspect of a given Muslim's life, which she manages and negotiates in accordance to other co-existing and sometimes contradictory moral and ethical registers. The latter view, however, privileges Islam as the central binding textual tradition that informs and gives rise to social forms and behaviour. I have sympathies with both positions. My experience in Luton taught me that Islam is indeed the most important identity marker for my informants. Muslims there *were* striving for coherence in their religious practices and outlooks. The Qur'an was a guiding text to both those who were familiar with it and those who were less so. The Prophet Mohammed was regarded as the exemplar of appropriate social and spiritual conduct. Yet my informants were only too aware that coherence and discipline were always ever fleeting in the face of so many other arresting social and personal pressures. This ranged from keeping up daily rituals (such as *salat*) and avoiding 'free mixing' with the opposite sex, or socialising with non-Muslim friends and colleagues. Even though so many of them persistently 'lapsed' in the 'proper' observation of religious conduct, very few were ultimately dissuaded from the pursuit. All of my friends in Luton wanted to be a 'good Muslim' one day. In this sense, Asad's insights are invaluable. Muslims in my fieldsite *did* consider Islam to be a unifying tradition that should inform every facet of their lives. I heard the phrase 'Islam is a way of life' so many times during my fieldwork from so many varying sources, young and old, pious and sinful.

In addition to representing a moral and spiritual guide, Islam was also a political identity to my informants. They shared a common emotional attachment to co-religionists both in their immediate vicinity and beyond.[3] Many cited global events affecting Muslims to be the driving force behind such affinities. Others pointed towards perceptions of Muslims in the media in the aftermath of 9/11 and 7/7. My older informants, who were familiar with popular and institutional racism growing up in the 1980s and 1990s argued that only Islam provided an alternative to divisive ideologies, such as ethnic nationalism. The idea of the *ummah*, therefore, was a concrete one, albeit for different reasons, depending on the person. My informants all bought into the notion that they were equal in front of God. If the community bled anywhere in the world, so did they. This powerful political outlook sat side by side with a sense of individuated Britishness. British-born Muslims did not identify with their countries of origin beyond the spectre of ancestral loyalty and pride. Some had visited South Asia on school holidays, for family weddings or other formal occasions. Given the choice, however, they preferred partying holidays in Ibiza or pilgrimages to Mecca. It is this paradox that is central to identity production among my informants. I have suggested throughout this book that transnational migration and the specific conditions of settlement and subsequent generational renewal has led to the production of such complex intersectional identities. With these interventions in mind, I ask the question: 'where does the anthropology of Islam sit in relation to postcolonial diasporas, minority citizenship and stigmatised masculinities?'

Anthropologists of Islam should consider moving beyond the nexus of observing discursive piety and associated incoherencies in majoritarian Muslim societies. A more sustained effort must be made to explore nebulous and transnational social forms that develop outside of the classical Islamic world. Muslims in diaspora, particularly those living in the

West, face a unique set of sociological challenges that inform self-making processes, which are otherwise absent from the ethnographic gaze in the Islamic heartlands. This absence may be in some way linked to the epistemological development of anthropology as a discipline, with its genealogical and territorial emphasis on the exotic Other.[4] It may also be due to the historic tensions between the 'worlds' of Christendom and Islam, and relations of dominance and subordination representative of colonial encounters that have, in turn, manifested in fetishised Orientalist scholarship.[5] Whatever the reasons, the events of 9/11 proved to represent a watershed moment not just in terms of a shift in global polity but also in the rendering of academic attention to a new generation of Muslims who have been forced to reconcile their background, beliefs and loyalties in the face of ever-increasing hegemonic scrutiny from non-Muslim observers and commentators. This is nowhere more acute than within diaspora contexts in the West.

Second- and third-generation Muslims find themselves in a peculiar situation where they are stigmatised in their own countries for their cultural and religious background.[6] This '9/11 generation'[7] has had to contend with wider 'moral panics', critical of conceptions of honour, patriarchal attitudes and the oppression of Muslim women.[8] Muslim men are viewed as hyper-masculine 'folk devils' prone to acts of devastating violence and antipathy towards Western liberal 'values'.[9] Although, from my experience, patriarchal dominance did exist to worrying degrees in Luton and there was a disproportionate involvement of Muslim men in crime and gang-related violence, ethnographic scrutiny suggests that Muslims, even in such a relatively small community as Luton, cannot be described or analysed as the homogenous group that these wider discourses perpetuate. This is also not lost on my informants, where many of them cited negative stereotyping of Muslims as terrorists and abusers of women as a reason to further uphold their Muslim identity in defiance of perceived injustice.[10] This particular set of temporal and material circumstances, I believe, adds further nuance and sophistication to the anthropological study of Islam even in majority-Muslim countries. In an ever-globalising world, to what extent are Muslims really separated by national borders and imagined civilizational barriers? In an era where Islamic discourses are transmitted through de-territorialised spaces, how long can vernacularised local forms of Islam resist transnational influences? And what types of Islamic expression flourish within socio-economically marginalised immigrant communities?

This book has attempted to familiarise the reader with the distinctive cultural habitus that my informants were born and raised within, much of which was constructed by pioneer post-war migrants as settlement in Britain became realised. Mostly escaping poverty and social upheaval in their home regions, migrants sought protection and solace within their own ethnic or national groups in the face of widespread racial and cultural hostility. In the preliminary phase of migration, South Asian village and clan networks facilitated ease of passage and the subsequent settlement for migrant men. With the arrival of wives and children, however, these networks were innovated to loosely resemble those in operation in their South Asian villages of origin. Migrants not only re-worked these networks in an effort to combat hostility from the host society but also as a means to provide social and economic stability in an unfamiliar land. The re-application of *biradari* networks, for example, ensured economic protection and provided a means by which first-generation parents could transfer South Asian cultural knowledge to their children. Thus, crucially, continuing its reproduction within diaspora. Muslim parents I spoke with commonly cited wanting to protect their children from the perceived corruption and moral malaise of wider British society.

The ensuing creation of distinct Muslim enclaves in British towns and cities, that exist to the present day, bears testament to this process. In these areas, mosques, *halal* butchers, Muslim-owned clothing and food establishments, and the sight of Muslims going about their daily business cannot escape notice. Luton is one such place. Young Muslims there were raised within this inimitable landscape. I have argued that British-born Muslims in Luton strategically conform to the conventions and protocols of the home and community in varied ways.[11] Although many are now critical of the application of South Asian 'culture' in Britain, a potent sense of respect and cooperation towards parents' wishes and aspirations was generally maintained. However, this is not to suggest that they did not attempt to persuade parents and community elders to compromise on certain issues. This was particularly pronounced on the issue of marriage, for example.

Despite being exposed to wider society in a more concerted way compared with previous generations, young Muslims in Luton predominantly continued to work and socialise with fellow Muslims. This does not necessarily suggest that they were antagonistic towards non-Muslims – many of my informants, particularly those who went to university, maintained social and professional links with non-Muslims. Rather that, in their everyday lives, broad access to non-Muslims was restricted due to work commitments and a desire to continue social relationships forged during childhood and adolescence. These relationships were usually established with other Muslims who lived and worked in and around the Bury Park area. Furthermore, my informants were obliged to financially contribute to the family once they reached employable age. Some of them were employed within family businesses, and spent long hours at work for little or no pay. Those who held managerial positions in large companies often supported friends and family members by securing employment for them within their companies. Some went as far as delving into organised crime with older brothers and cousins, the profits from which were remitted to parents.

Additionally, being Muslim provided a sense of solidarity and emotive belonging situated in opposition to wider social profiling.[12] As postcolonial theorists such as Homi Bhabha (1994), Stuart Hall (2000) and Paul Gilroy (1994) remind us, diaspora identities are strategic choices made by actors in order to compensate for material and discursive realities. Moreover, they suggest that subjectivities are contingent to relations of power that operate *within* and *without* the individual and the nation state.[13] Much like the pioneer generation, British-born Muslims continue to live in a society that remains seemingly hostile to cultural and religious difference.[14] Many of my informants felt that the nation did not include them within Benedict Anderson's 'imagined community' (1991). It is here that the idea of the *ummah* was particularly efficacious for my informants. Post-9/11 discourses and the 'war on terror' has fuelled the feeling that Muslims are the *de facto* 'Other', not just in Britain, but across the globe.

This book has aimed to ethnographically explore the complex identities emerging among British-born Muslim men in Luton. My informants in this controversial English town sought respite in an overt yet multifarious 'Islamic' identity that attempted to reconcile their sense of perceived 'difference'. The rejection of official narratives of the nation and the inadequacies of South Asian 'village culture' was consciously negotiated on an everyday basis. Through this process, young Muslims were developing distinct articulations of the 'self' that departed from hegemonic understandings of Muslims as a homogenised group. Through their refashioning of what it means to be Muslim *and* British, my informants in Luton demonstrated that diaspora identities are always in motion, historically determined and culturally hybrid.

Notes

1. See Wikan 2002; Silverstein 2004, Ewing 2008; Hopkins 2008; Werbner 2002; Mandel 2008; and Jouili 2015.
2. See Schielke 2009; Marsden and Retsikas 2013; and Masquelier and Soares 2016.
3. Compare with Marranci 2008.

4. See Kuper 1983; and Stocking 1991.
5. See Asad 1973; Said 1978; Abu-Lughod 1989; Mamdani 2004; and compare with Lewis 1990; Huntington 1996.
6. See Ewing 2008.
7. See Maira 2009, 2016; Masquelier and Soares 2016; and Herrera and Bayat 2010.
8. See Archer 2003.
9. See Alexander 2000.
10. Compare with Herzfeld 1985.
11. See DeHanas 2016; and compare with Jouili 2009.
12. Compare with Gardner and Shukur 1994.
13. See Foucault 2003.
14. Compare with Ewing 2008.

References

Abbas, T. (2011) *Islamic Radicalism and Multicultural Politics: The British Experience*. London, New York: Routledge.

Abedin, S. Z. and Z. Sardar (eds) (1995) *Muslim Minorities in the West*. London: Grey Seal.

Abu-Lughod, L. (1989) 'Zones of Theory in the Anthropology of the Arab World'. *Annual Review of Anthropology*, 18: 267–306.

Abu-Lughod, L. (2002) 'Do Muslim Women Really Need Saving? Anthropological Reflections, Cultural Relativism and its Others'. *American Anthropologist*, 104(3): 783–90.

Adams, C. (1987) *Across Seven Seas and Thirteen Rivers: Life Stories of Pioneer Sylheti Settlers in Britain*. London: Tower Hamlets Arts Project (THAP) Books.

Ahmad, F. (2006) 'British Muslim Perceptions and Opinions on News Coverage of September 11'. *Journal of Ethnic and Migration Studies*, 32(6) (August 2006): 961–82.

Ahmed, F. and M. S. Seddon (eds) (2012) *Muslim Youth: Challenges, Opportunities, and Expectations*. London, New York: Continuum.

Ahmed, T. S. (2005) 'Reading between the Lines – Muslims and the Media' in Abbas, Tahir (ed.) *Muslim Britain: Communities Under Pressure*. London: Zed Books.

Alam, M. Y. (ed.) (2006) *Made in Bradford*. Pontefract: Route Publishing.

Alavi, H. A. (1972) 'Kinship in West Punjab Villages'. *Contributions to Indian Sociology*, 6:1.

Alexander, C. E. (1996) *The Art of Being Black: The Creation of Black British Youth*. Oxford: Clarendon Press.

Alexander, C. E. (2000) *The Asian Gang: Ethnicity, Identity, Masculinity*. Oxford, New York: Berg.

Alexander, C. E., V. Redclift and A. Hussain (eds) (2013) *The New Muslims*. London: Runnymede Trust.

Ali, J. (2003) 'Islamic Revivalism: The Case of Tablighi Jamaat'. *Journal of Muslim Minority Affairs*, 23(1) (April 2003): 173–80.

Allen, C. (2005) 'From Race to Religion: the New Face of Discrimination' in Abbas, T. (ed.) *Muslim Britain: Communities Under Pressure*. London: Zed Books.

Allen, C. (2010) *Islamophobia*. London, New York: Routledge.

Allen, C. (2012) 'A part of the fabric of everyday life: Islamophobia in contemporary Britain' in Siddiqui, A. and Lamb, C. (eds) *Meeting Muslims*. Leicester: Christians Aware.

Allen, G. and C. Watson (2017) *UK Prison Population Statistics*. House of Commons Library Briefing Paper, 20 April 2017.

Allen, S. (1971) *New Minorities, Old Conflicts*. London: Random House.

Alleyne, B. (2002) 'An idea of community and its discontents: towards a more reflexive sense of belonging in Multicultural Britain'. *Ethnic and Racial Studies*, 25(4): 607–27.

Anderson, B. (1991) *Imagined Communities: reflections on the origins and spread of nationalism*. London, New York: Verso.

Ansari, H. (2004) *The Infidel Within: Muslims in Britain since 1800*. London: Hurst & Co.

Anwar, M. (1976) 'Young Asians between two cultures'. *New Society*, 38(16): 563–65.

Anwar, M. (1979) *The Myth of Return: Pakistanis in Britain*. London: Heinemann.

Anwar, M. (1998) *Between Cultures: Continuity and Change in the Lives of Young Asians*. London: Routledge.

Appadurai, A. (1995) 'The Production of Locality' in Fardon, R. (ed.) *Counterworks: Managing the Diversity of Knowledge*. London, New York: Routledge.

Appadurai, A. (1996) *Modernity at Large: Cultural Dimensions of Globalization*. Minneapolis, London: University of Minnesota Press.

Archer, L. (2003) *Race, Masculinity and Schooling: Muslim boys and education*. Maidenhead: Open University Press.

Archer, T. (2009). 'Welcome to the umma: The British state and its Muslim citizens since 9/11'. *Cooperation and Conflict*, 44(3):329–47.

Asad, T. (1973) *Anthropology and the Colonial Encounter*. Reading: Ithaca Press.

Asad, T. (1986) 'The Idea of an Anthropology of Islam'. *Occasional Paper Series* 1–17, Georgetown University.

Awan, I. (2012) '"I Am a Muslim Not an Extremist": How the Prevent Strategy Has Constructed a "Suspect" Community'. *Politics and Polity*, 40(6): 1158–85.

Back, L. (1993) 'Race, identity and nation within an adolescent community in South London'. *New Community*, 19(2) (January 1993): 217–33.

Back, L. (1996) *New Ethnicities and Urban Culture: racism and multiculture in young lives*. London, New York: Routledge.

Baker-Beall, C., C. Heath-Kelly, and L. Jarvis (2015) *Counter-Radicalisation: Critical Perspectives*, Abingdon: Routledge

Ballard, R. (1988) 'The Political Economy of Migration: Pakistan, Britain and the Middle East', in Eade, J. (ed.), *Migrants, Workers, and the Social Order*. London: Tavistock.

Ballard, R. (1990) 'Migration and Kinship: the differential effect of marriage rules on the processes of Punjabi migration to Britain' in Clarke, C., C. Peach and S. Vertovec (eds) *South Asians Overseas: Migration and Ethnicity*. Cambridge: Cambridge University Press.

Ballard, R. (ed.) (1994) *Desh Pardesh: the South Asian Presence in Britain*. London: Hurst & Co.

Banerji, S. and Baumann, G. (1990) 'Bhangra 1984–8: Fusion and Professionalization in the Genre of South Asian Dance Music' in Oliver, Paul (ed.) *Black Music in Britain: essays on the Afro-Asian contribution to popular music*. Maidenhead: Open University Press.

Bartlett J. and M. Littler (2011) *Inside the EDL: populist politics in a digital age*. London: Demos.

Basch, L., N. G. Schiller and C. S. Blanc (1994) *Nations Unbound: Transnational Projects, Postcolonial Predicaments and Deterritorialized Nation-States*. London, New York: Routledge.

Baumann, G. (1996) *Contesting Culture: Discourses of identity in multi-ethnic London*. Cambridge: Cambridge University Press.

Bayart, A. (2007) 'Islamism and the Politics of Fun'. *Public Culture*, 19(3): 433–59.

Becher, H. (2008) *Family Practices in South Asian Muslim Families: Parenting in a Multi-Faith Britain*. London: Palgrave Macmillan.

Beckles, C. A. (1998) '"We Shall Not Be Terrorised Out of Existence": The Political Legacy of England's Black Bookshops'. *Journal of Black Studies*, 29(1) (September 1998): 51–72.

Benyon, J. (ed.) (1984) *Scarman and After: Essays Reflecting on Lord Scarman's Report, the Riots and Their Aftermath*. London: Pergamon Press.

Bhabha, H. K. (1994) *The Location of Culture*. London, New York: Routledge.

Bhachu, P. (1985) *Twice Migrants: East African Sikh Settlers in Britain*: London: Tavistock.

Bhachu, P. (1991) 'Culture, ethnicity and class among Punjabi Sikh women in 1990s Britain'. *New Community*, 17(3) (April 1991): 401–12.

Black, L. (1993) 'Race, identity and nation within an adolescent community in South London'. *New Community*, 19(2): 217–33.

Bledsoe, Caroline H. and Robey, Kenneth M. (1993) 'Arabic Literacy and Secrecy among the Mende of Sierra Leone' in Street, Brian V. (ed.) *Cross-Cultural Approaches to Literacy*. Cambridge: Cambridge University Press.

Blunt, E. A. H., (1931) *The Caste System of Northern India*. Oxford: Oxford University Press.

Bowen, J. R. (2004) 'Beyond Migration: Islam as a Transnational Public Space'. *Journal of Ethnic and Migration Studies*, 30(5) (September 2004): 879–94.

Bowen, J. R. (2011) 'How could English courts recognize Shariah?' *University of St. Thomas Law Journal*, 7: 411–35.

Bowen, J. R. (2012) *A New Anthropology of Islam*. Cambridge: Cambridge University Press.

Bowen, J. R. (2013) 'Sanctity and shariah: Two Islamic modes of resolving disputes in today's England' in F. von Benda-Beckmann, K. von Benda-Beckmann, M. Ramstedt and B. Turner (eds) *Religion in disputes*. New York: Palgrave.

Bowen, J. R. (2016) *On British Islam: Religion, Law, and Everyday Practice in Shari'a Councils*. Princeton, NJ; London: Princeton University Press.

Breen-Smyth, M (2014) 'Theorising the "Suspect Community": counterterrorism, security practises and the public imagination'. *Critical Studies on Terrorism*, 7(2): 223–40.

Brenner, L. (2001) *Controlling Knowledge: Religion, Power and Schooling in a West African Muslim Society*. Bloomington: Indiana University Press.

Brown, K. E. (2010) 'Contesting the Securitization of British Muslims: Citizenship and Resistance'. *Interventions*, 12(2):171–82.

Bunglawala, I. (2002) 'British Muslims and the Media' in *The Quest for Sanity: Reflections on September 11 and the Aftermath*. Muslim Council of Britain.

Byron, M. (1994) *Post-War Caribbean Migration to Britain: the unfinished cycle*. Aldershot: Ashgate.

Castles, S. (1987) *Here for Good: Western Europe's new ethnic minorities*. London: Pluto Press.

Castles, S. and Koszak, G. (1973) *Immigrant Workers and Class Structure in Western Europe*. Oxford: Oxford University Press.

Chih, Rashida (2007) 'What is a Sufi Order? Revisiting the concept through a case study of the Khalwatiyya in contemporary Egypt' in M. Van Bruinessen and J. Day Howell (eds) *Sufism and the Modern*. London: I.B. Tauris.

Choudhury, T. and H. Fenwick (2011) 'The Impact of counter-terrorism measures on Muslim communities'. *International Review of Law, Computers & Technology*, 25(3) November 2011: 151–81.

Clarke, C., C, Peach, and S. Vertovec (1990) *South Asians Overseas: migration and ethnicity*. Cambridge: Cambridge University Press.

Considine, C. (2018) *Islam, Race, and Pluralism in the Pakistani Diaspora*. London, New York: Routledge.

Cornwall, A. and N. Lindisfarne (eds) (1994) *Dislocating Masculinity: Comparative Ethnographies*. London, New York: Routledge.

Cornwall, A., F. G. Karioris and N. Lindisfarne (eds) (2016) *Masculinities Under Neoliberalism*. London: Zed Books.

Cressey, G. (2006) *Diaspora Youth and the Ancestral Homeland: British Pakisani/Kashmiri Youth visiting kin in Pakistan and Kashmir*. Leiden: Brill.

Cunningham, S. and J. Sinclair (2000) *Floating Lives: The Media and Asian Diaspora – Negotiating Cultural Identity through Media*. St Lucia: Queensland University Press.

Dahya, B. (1973) 'Pakistanis in Britain: Transients or Settlers?'. *Race & Class*, 14(3): 241–77.

Das, V. (1984) 'For a folk-theology and theological anthropology of Islam'. Contributions of Indian Sociology, 18(2): 293–300.

De Certeau, M. (1984) The Practice of Everyday Life. Berkeley, Los Angeles: University of California Press.

Deeb, L. (2006). 'An Enchanted Modern: Gender and Public Piety in Shi'i Lebanon'. Princeton, NJ: Princeton University Press, 111–13.

Deeb, L. (2009) 'Emulating and/or embodying the ideal: The gendering of temporal frameworks and Islamic role models in Shi'i Lebanon'. *American Ethnologist*, 36(2): 242–57.

Deeb, L. and M. Harb (2017) *Leisurely Islam: Negotiating Geography and Morality in Shi'ite South Beirut*. Princeton, NJ, and Oxford: Princeton University Press.

DeHanas D. N. (2016) *London Youth, Religion, and Politics: Engagement and Activism from Brixton to Brick Lane*. Oxford: Oxford University Press.

Din, I. (2006) *The New British: the impact of culture and community on young Pakistanis*. Aldershot: Ashgate.

Douglas, G., Doe, N., Gilliat-Ray, S., Sandberg, R. and Khan, A. (2011) 'Social Cohesion and Civil Law: Marriage, Divorce and Religious Courts'. *SSRN Electronic Journal*. http://www.law.cf.ac.uk/clr/Social%20Cohesion%20and%20Civil%20Law%20Full%20Report.pdf

Douglas, G., Doe, N., Gilliat-Ray, S., Sandberg, R. and Khan, A. (2012) 'The role of religious tribunals in regulating marriage and divorce'. *Child and Family Law Quarterly*, 24(2): 139–57.

Eade, J. (1989) *The Politics of Community: the Bangladeshi community in East London*. Aldershot: Avebury.

Eade, J. (1996) 'Nationalism, Community and the Islamization of Space' in B. D. Metcalf (ed.) *Making Muslim Space in North America and Europe*. Berkeley, Los Angeles, London: University of California Press.

Eade, J. (ed.) (1987) *Migrants, Workers and the Social Order*. London: Tavistock.

Eikelman D. F. and J. Piscatori (1996) *Muslim Politics*. Princeton, NJ: Princeton University Press.

Elgar, Z. (1960) *A Punjabi Village in Pakistan*. New York: Colombia University Press.

El-Zein, A.H. (1977) 'Beyond Ideology and Theology: The Search for an Anthropology of Islam'. *Annual Review of Anthropology*, 6: 227–54.

Ewing, K. P. (1990) 'The Illusion of Wholeness: Culture, Self, and the Experience of Inconsistency'. *Ethos*, 18(3): 251–78.

Ewing, K. P. (2008) *Stolen Honour: Stigmatizing Muslim Men in Berlin*. Stanford, CA: Stanford University Press.

Ewing, K. P. (ed.) (2008) *Being and Belonging: Muslims in the United States since 9/11*. New York: Russell Sage Foundation.

Fadil, N. (2009) 'Managing Affects and Sensibilities: The Case of Not-Handshaking and Not-Fasting'. *Social Anthropology*, 17(4): 439–54.

Fanon, F. (1967) *The Wretched of the Earth*. London: Penguin Books.

Fernando, M. L. (2016) 'The Unpredictable Imagination of Muslim French: Citizenship, Public Religiosity, and Political Possibility in France' in A. Masquelier and B. Soares (eds) *Muslim Youth and the 9/11 Generation*. Santa Fe: University of Mexico Press.

Foreman, C. (1989) *Spitalfields: a battle for land*. London: Hilary Shipman.

Foucault, M. (1986) 'Truth and Power' and 'Power and Strategies' in C. Gordon (ed.) *Power/Knowledge: Selected Interviews and Other Writings 1972–1977*. Hemel Hempstead: The Harvester Press.

Foucault, M. (1989) The Archaeology of Knowledge. London, New York: Routledge.

Foucault, M. (2003) 'The Subject and Power' in P. Rabinow and N. Rose (eds) *The Essential Foucault; Selections from Essential Works of Foucault 1954–1984*. New York: The New Press.

Franceschelli, M. (2016) *Identity and Upbringing in South Asian Muslim Families: Insights from Young People and their Parents in Britain*. London: Palgrave Macmillan.

Fryer, P. (1984) *Staying Power: The history of black people in Britain*. London: Pluto Press.

Gardner, K. (1993) 'Desh-Bidesh: Sylheti Images of Home and Away'. *Man*, New Series, 28(1) (March 1993): 1–15.

Gardner, K. (1995) *Global Migrants, Local Lives: Travel and Transformation in Rural Bangladesh*. Oxford: Clarendon Press.

Gardner, K. (2002) *Age, Narrative and Migration: The Life Course and Life Histories of Bengali Elders in London*. Oxford: Berg.

Gardner, K. and A. Shukur (1994) 'I'm Bengali, I'm Asian, and I'm Living Here: The Changing Identity of British Bengalis' in R. Ballard (ed.) *Desh Pardesh: the South Asian Presence in Britain*. London: Hurst & Co.

Garnett, J. and S. L. Hausner (eds) (2015) *Religion in Diaspora: Cultures of Citizenship*. London: Palgrave Macmillan.

Geertz, C. (1971) *Islam Observed: Religious Development in Morocco and Indonesia*. Chicago: University of Chicago Press.

Geertz, C. (1973) *The Interpretation of Cultures*. Basic Books.

Gellner, E. (1981) *Muslim Society*. Cambridge: University of Cambridge Press.

Gellner, E. (1983) Nations and Nationalism. Malden, Oxford: Blackwell.

Gest, J. (2010) *Apart: Alienated and Engaged Muslims in the West*. London: Hurst & Co.

Gilroy, P. (1987) *There Ain't No Black in the Union Jack: the cultural politics of race and nation*. London, New York: Routledge.

Gilroy, P. (2005) 'Multiculture, double consciousness and the "war on terror"'. *Patterns of Prejudice*, 39(4): 431–43.

Gilroy, P. (2006) *Postcolonial Melancholia*. Columbia University Press.

Gilsenan, M. (1982) *Recognizing Islam*. London: Croom Helm.

Githens-Mazer, J. (2012) 'The rhetoric and reality: Radicalisation and political discourse'. *International Political Science Review*, 33(5): 556–67.

Glynn, S. (2002) 'Bengali Muslims: the new East End radicals?' *Ethnic and Racial Studies*, 25(6): 969–88.

Glynn, S. (2004) 'East End Immigrants and the Battle for Housing: a comparative study of political mobilisation in the Jewish and Bengali communities'. Final version published in the *Journal of Historical Geography*, 31: 528–45 (2005).

Gole, N. (2015) *The Daily Lives of Muslims*. London: Zed Books.

Grillo, R. D. (2003) 'Cultural essentialism and cultural anxiety'. *Anthropological Theory*, 3(2): 157–73.

Grillo, R. D. (2004) 'Islam and Transnationalism'. *Journal of Ethnic and Migration Studies*, 30(5) (September 2004): 861–78.

Grillo, R. D. (2018) *Transnational Migration and Multiculturalism: Living with Difference in a Globalised World*. Lewes: B & RG Books.

Gupta A. and J. Ferguson (1997) *Culture, Power, Place: Explorations in Critical Anthropology*. Durham, NC: Duke University Press.

Hall, K. D. (2002) *Lives in Translation: Sikh Youth as British Citizens*. Philadelphia: University of Pennsylvania Press.

Hall, S. (1993) 'Culture, Community, Nation'. *Cultural Studies*, 7(3): 349–63.

Hall, S. (2000) 'The Multicultural Question' in B. Hesse (ed.) *Unsettled Multiculturalisms: Diasporas, Entanglements*. London: Zed Books.

Hall, S. (2003) 'Cultural Identity and Diaspora' in J. E. Braziel and A. Mannur (eds) *Theorizing Diaspora*. Malden, Oxford: Blackwell.

Hamid, S. (2009) 'The Attraction of "Authentic" Islam: Salafism and British Muslim Youth' in R. Meijier (ed.) *Global Salafism: Islam's New Religious Movement*. New York: Colombia University Press.

Hamid, S. (2016) *Sufis, Salafis and Islamists: The Contested Ground of British Islamic Activism*. London, New York: I.B. Tauris.

Heath-Kelly, C. (2013) 'Counter-Terrorism and the Counterfactual: Producing the "Radicalisation" Discourse and the UK PREVENT Strategy'. *British Journal of Politics and International Relations*, 15: 394–415.

Henkel, H. (2005). '"Between belief and unbelief lies the performance of salaat": Meaning and efficacy of a Muslim ritual'. *Journal of the Royal Anthropological Institute*, 11(3) (September): 487–507.

Herrera, L. and A. Bayat (eds) (2010) *Being Young and Muslim: New Cultural Politics in the Global South and North*. Oxford, New York: Oxford University Press.

Herzfeld, M. (1985) *The Poetics of Manhood: Contest and Identity in a Cretan Mountain Village*. Princeton, NJ: Princeton University Press.

Hinnells, J. R. (2007) *Religious Reconstruction in the South Asian Diasporas: From One Generation to Another*. London: Palgrave Macmillan.

Hirschkind, C. (2001) 'The Ethics of Listening: Cassette-Sermon Audition in Contemporary Egypt'. *American Ethnologist* 28(3): 623–49.

Holden, L. (2003) *Vauxhall Motors and the Luton Economy, 1900–2002*. Bedfordshire Historical Record Society. Woodbridge: Boydell Press.

Hopkins, P. (2008) *The Issue of Masculine Identities for British Muslims after 9/11*. New York: The Edwin Mellen Press.

Hopkins, P. and R. Gale (eds) (2009) *Muslims in Britain: Race, Place, and Identities*. Edinburgh: Edinburgh University Press.

Hoque, A. (2015) 'Muslim Men in Luton, UK: "Eat First, Talk Later"'. *South Asia Research*, 35(1): 81–102.

Horstman, A. (2007) 'The Inculturation of Transnational Islamic Missionary Movement: Tablighi Jamaat al-Daw and Muslim Society in Southern Thailand'. *Sojourn: Journal of Social Issues in Southeast Asia*, 22(1): 107–30.

Huntington, S. P. (1996) *The Clash of Civilisations and the Remaking of World Order*. New York: Simon and Schuster.

Hussain, S. (2008) *Muslims on the Map: A National Survey of Social Trends in Britain*. London, New York: Tauris Academic Studies.

Hutton, J. H. (1946) *Caste in India*. Oxford: Oxford University Press.

Inge, A. (2017) *The Making of a Salafi Muslim woman: Paths to Conversion*. Oxford: Oxford University Press.

Jacobson, J. (1998) *Islam in Transition: Religion and Identity among British Pakistani Youth*. London, New York: Routledge.

Janmohamed, S. (2016) *Generation M: Young Muslims Changing the World*. London, New York: I.B. Tauris.

Janson, Marloes (2014) *Islam, Youth, and Modernity in the Gambia: The Tablighi Jama'at*. Cambridge: Cambridge University Press.

Jarvis, L., and Lister, M. (2013) 'Disconnected Citizenship? The Impacts of Anti-terrorism Policy on Citizenship in the UK'. *Political Studies*, 61(3): 656–75.

Jeffery, Patricia (1976) *Migrants and Refugees: Muslim and Christian Pakistani Families in Bristol*; Cambridge: Cambridge University Press.

Jeffrey, C. and J. Dyson (eds) (2008) *Telling Young Lives: portraits of global youth*. Philadelphia, PA: Temple University Press.

Johnson, P. C. (2012) 'Religion and Diaspora'. *Religion and Society* 3: 27–9.

Joppke, C. (2009) 'Limits of Integration Policy: Britain and her Muslims'. *Journal of Ethnic and Migration Studies*, 35(3) (March 2009): 453–72.

Jouili, J.S. (2009) 'Negotiating Secular Boundaries: Pious Micro-Practices of Muslim Women in French and German Public Spheres'. *Social Anthropology*, 17(4):455–70.

Kabir, N. A. (2010) *Young British Muslims: Identity, Culture, Politics, and the Media*. Edinburgh: Edinbugh University Press.

Kabir, N. A. (2013) *Young American Muslims: Dynamics of Identity*. Edinburgh: Edinbugh University Press.

Kahani-Hopkins, V. and N. Hopkins (2002) '"Representing" British Muslims: the strategic dimension to identity construction'. *Ethnic and Racial Studies*, 25(2): 288–309.

Kapoor, I. (2002). 'Capitalism, culture, agency: Dependency versus postcolonial theory'. *Third World Quarterly*, 23(4): 647–64.

Kibria, N. (2008) 'The 'New Islam' and Bangladeshi youth in Britain and the US'. *Ethnic and Racial Studies*, 31(2): 243–66.

Kibria, N. (2011) *Muslims in Motion: Islam and National Identity in the Bangladeshi Diaspora*. London: Rutgers University Press.

Kundnani, A. (2001) 'From Oldham to Bradford: the violence of the violated'. *Race and Class*, 43: 105–10.

Kundnani, A. (2002) 'The Death of Multiculturalism'. *Race and Class*, 43: 67–72.

Kundnani, A. (2007) 'Integrationism: the politics of anti-Muslim racism'. *Race and Class*, 48(4): 24–44.

Kundnani, A. (2012) 'Radicalisation: the journey of a concept'. *Race and Class*, 54(2): 3–25.

Kundnani, A. (2014) *The Muslims Are Coming! Islamophobia, Extremism, and the Domestic War on Terror*. London, New York: Verso.

Kuper, A. (1983) *Anthropology and Anthropologists: The Modern British School*. London, New York: Routledge.

Lambek, M. (ed.) (2010) *Ordinary Ethics: Anthropology, Language, and Action*. New York: Fordham University Press.

Layton-Henry, Z. (1992) *The Politics of Immigration: Immigration, 'Race' and 'Race' Relations in Post-war Britain*. Malden, Oxford: Blackwell.

Lewis, B. (1990) 'The Roots of Muslim Rage'. *The Atlantic*, 266(3): 47–60.

Lewis, Philip (1994) *Islamic Britain: Religion, Politics and Identity among British Muslims*; London: I.B. Tauris.

Lewis, Philip (2007) *Young, British and Muslim*. London, New York: Continuum.

Linke, U. (2014) 'Racializing cities, naturalizing space: The seductive appeal of iconicities of dispossession'. *Antipode*, 46(5): 1222–39.

Lister, M. and L. Jarvis (2013) 'Disconnection and Resistance: Anti-terrorism and Citizenship in the UK'. *Citizenship Studies*, 17(6–7): 756–69.

Lyon, S. M. (2004) *An Anthropological Analysis of Local Politics and Patronage in a Pakistani Village*. New York: The Edwin Mellen Press.

Mahmood, S. (2001) 'Rehearsed Spontaneity and the Conventionality of Ritual: Disciplines of Ṣalat'. *American Ethnologist* 28(4): 827–53.

Mahmood, S. (2005) *The Politics of Piety: The Islamic Revival and the Feminist Project*. Princeton, NJ: Princeton University Press.

Maira, S. M. (2009) *Missing: Youth, Citizenship, and Empire after 9/11*. Durham, NC, London: Duke University Press.

Maira, S. M. (2016) *The 9/11 Generation: Youth, Rights, and Solidarity in the War on Terror*. New York: New York University Press.

Maira, S. M. and E. Soep (eds) (2005) *Youthscapes: The Popular, the National, the Global*. Philadelphia: University of Pennsylvania Press.

Malik, J. (2006) 'Madrasah in South Asia' in Ibrahim, A. M. (ed.) *The Blackwell Companion to Contemporary Islamic Thought*. Basingstoke: Blackwell.

Mamdani, M. (2004) *Good Muslim, Bad Muslim: Islam, the USA and the global war against terror*. Permanent Black.

Mandaville, P. (2001) 'Reimagining Islam in Diaspora: The Politics of Mediated Community'. *The International Communication Gazette*, 63(2–3): 169–86.

Mandaville, P. (2004) *Transnational Muslim Politics: Reimagining the Umma*. London: Routledge.

Mandaville, P. (2009) 'Muslim Transnational Identity and State Responses in Europe and the UK after 9/11: Political Community, Ideology and Authority'. *Journal of Ethnic and Migration Studies*, 35(3): 491–506.

Mandel, R. (2008) *Cosmopolitan Anxieties: Turkish Challenges to Citizenship and Belonging in Germany.* Durham, NC: Duke University Press.

Marranci, G. (2008) *The Anthropology of Islam.* Oxford, New York: Berg.

Marsden, M. (2005) *Living Islam: Muslim Religious Experience in Pakistan's North-West Frontier.* Cambridge: Cambridge University Press.

Marsden, M. (2016) *Trading Worlds: Afghan Merchants Across Modern Frontiers.* London: Hurst & Co.

Marsden, M. and K. Retsikas (eds) (2013) *Articulating Islam: Anthropological Approaches to Muslim Worlds.* New York, London: Springer.

Masquelier, A. and B. Soares (2016) *Muslim Youth and the 9/11 Generation.* Santa Fe: University of Mexico Press.

McLoughlin, S. (2013) 'Imagining a Muslim Diaspora in Britain? Islamic Consciousness and Homelands Old and New' in Alexander, C. E., V. Redclift and A. Hussain (eds) *The New Muslims.* London: Runnymede Trust.

Meer, N. (2006) 'GET OFF YOUR KNEES: Print media public intellectuals and Muslims in Britain'. *Journalism Studies*, 4(1): 35–59.

Meer, N. (2008) 'The politics of voluntary and involuntary identities: are Muslims in Britain an ethnic, racial or religious minority?'. *Patterns of Prejudice*, 42(1): 61–81.

Meijer, R. (ed.) (2009) *Global Salafism: Islam's New Religious Movement.* New York: Colombia University Press.

Metcalf, B. D. (1982) *Islamic Revival in British India: Deoband, 1860–1900.* Oxford: Oxford University Press.

Miles, R. (1987) 'Recent Marxist Theories of Racism and the Issue of Racism'. *The British Journal of Sociology*, 38(1) (March 1987): 24–43.

Miles, Robert (1993) *Racism after 'race relations'.* London, New York: Routledge.

Mines, M. (1994) *Public Faces, Private Voices: Community and Individuality in South India.* Berkeley and Los Angeles: University of California Press.

Modood, T. (1990) *Muslims, race and equality in Britain: some post-Rushdie affair reflections.* Centre for the Study of Islam and Christian–Muslim Relations.

Modood, T. (2005) *Multicultural Politics: Racism, Ethnicity and Muslims in Britain.* Edinburgh: Edinburgh University Press.

Modood, T. (2007) *Multiculturalism.* Malden, Oxford: Polity Press.

Modood, T. (2009) 'Muslims and the Politics of Difference' in P. Hopkins and R. Gale (eds) *Muslims in Britain: Race, Place and Identities.* Edinburgh: Edinburgh University Press.

Modood, T. (ed.) (1997) *Ethnic Minorities in Britain: diversity and disadvantage.* London: Policy Studies Institute.

Modood T. and P. Werbner (eds) (1997) *The Politics of Multicultrualism in the New Europe: Racism, Identity and Community.* London, New York: Zed Books.

Mondal, A. A. (2008) *Young British Muslim Voices.* Santa Barbara: Greenwood World.

Morey, P. and A. Yaqin (2011) *Framing Muslims: Stereotyping and Representation after 9/11.* Cambridge, London: Harvard University Press.

Oakely, J. and H. Callaway (eds) (1992) *Anthropology and Autobiography.* London, New York: Routledge.

Osella, F. and B. Soares (eds) (2010) *Islam, Politics, Anthropology.* Oxford: Wiley-Blackwell.

Osella, F. and C. Osella (2000). 'Migration, Money and Masculinity in Kerala'. *The Journal of the Royal Anthropological Institute*, 6(1): 117–33.

Osella, F. and C. Osella (2013) *Islamic Reform in South Asia.* Delhi: Cambridge University Press.

Pandian, A. and D. Ali (eds) (2010) *Ethical Life in South Asia.* Bloomington: Indiana University Press.

Parekh, B. (2000) *Rethinking Multiculturalism: Cultural Diversity and Political Theory.* London: Palgrave Macmillan.

Patterson, S. (1963) *Dark Strangers.* London: Tavistock Publications.

Patterson, S. (1969) *Immigration and Race Relations in Britain 1960–1967.* Oxford: Oxford University Press.

Phizaklea, A. and Miles R. (1980) *Labour and Racism.* London: Routledge and Kegan Paul.

Pieri, Z. P. (2015) *Tablighi Jamaat and the Quest for the London Mega Mosque.* New York: Palgrave Macmillan.

Pilkington, H. (2016) *Loud and Proud: Passion and Politics in the English Defence League.* Manchester: Manchester University Press.

Pocock, D. F. (1976) 'Preservation of the religious life: Hindu immigrants in England'. *Contributions to Indian Sociology*, 10(2): 341–65.

Poole, E. (2002) *Reporting Islam: Media Representations of British Muslims*. London: I.B. Tauris.

Rabinow, P. (ed.) (1991) *The Foucault Reader: An Introduction to Foucault's Thoughts*. London, New York: Penguin Books.

Raza, M. S. (1991) *Islam in Britain: past, present and the future*. Volcano Press Ltd.

Robinson, F. (1983) 'Islam and Muslim Society in South Asia'. *Contributions in Indian Sociology*, 17(2):185–203.

Robinson, F. (2008) 'Islamic Reform and Modernities in South Asia'. *Modern Asian Studies* 42 (2/3): 259–81.

Roy, O. (2004) *Globalised Islam*. London: Hurst & Co.

Sageman, M. (2004) *Understanding Terror Networks*. Philadelphia: University of Pennsylvania Press.

Sageman, M. (2008) *Leaderless Jihad: terror networks in the twenty-first century*. Philadelphia: University of Pennsylvania Press.

Said, E. W. (1978) *Orientalism*. London, New York: Routledge and Kegan Paul.

Saifullah Khan, V. (1977) 'The Pakistanis: Mirpuri villagers at Home and in Bradford' in J. L. Watson (ed.) *Between Two Cultures: Migrants and Minorities in Britain*. Oxford: Basil Blackwell.

Saifullah Khan, V. (1979) 'Migration and Social Stress' in Saifullah Khan, V. (ed.) *Minority Families in Britain: Support and Stress*. Basingstoke: Macmillan.

Sanneh, Lamin (1997) *The Crown and the Turban: Muslims and West African Pluralism*. Boulder, CO: Westview Press.

Scarman, L. G. (1983) *The Scarman report: the Brixton disorders, 10–12 April 1981*. Harmondsworth, Middlesex: Penguin.

Schielke, S. (2009) 'Being Good in Ramadan: Ambivalence, Fragmentation, and the Moral Self in the Lives of Young Egyptians'. *Journal of the Royal Anthropological Institute* 15: S24–S40.

Schielke, S. (2010). 'Second thoughts about the anthropology of Islam, or how to make sense of grand schemes in everyday life'. *ZMO Working Papers*, 2: 1–16.

Schielke, S. and L. Debevec (eds) (2012) *Ordinary Lives and Grand Schemes: An Anthropology of Everyday Religion*. New York, Oxford: Berghahn.

Schimmel, A. (1975) *Mystical Dimensions of Islam*. Chapel Hill: University of North Carolina Press.

Scourfield, J., C. Taylor, G. Moore and S. Gilliat-Ray (2012) 'The Intergenerational Transmission of Islam in England and Wales: Evidence from the Citizenship Survey'. *Sociology* 46(1): 91–108.

Sedgwick, M. (2010) 'The concept of radicalization as a source of confusion', *Terrorism and Political Violence*, 22(4): 480–501.

Shah, P. (2013): 'In pursuit of the pagans: Muslim law in the English context'. *Journal of Legal Pluralism*, 45(1): 58–75.

Shaw, A. (1988) *A Pakistani Community in Britain*. Oxford: Basil Blackwell.

Shaw, A. (2000) *Kinship and Continuity: Pakistani Families in Britain*. London: Harwood.

Shaw, A. (2001) 'Kinship, Cultural Preference and Immigration: Consanguineous Marriage among British Pakistanis'. *Royal Anthropological Institute*, New Series, 7: 315–34.

Sheridan, Lorraine (2002) 'Religious Discrimination: The New Racism' in Hamid, A. W and J. Sherif (eds) *The Quest for Sanity: Reflections on September 11 and the Aftermath*. Muslim Council of Britain.

Shukra, K. (1998) *The Changing Patterns of Black Politics in Britain*. London: Pluto Press.

Sikand, Y. (2002) *The Origins and Development of Tablighi Jamaat (1920–2000): A Cross-country Comparative Study*. Hyderabad: Sangham Books.

Sikand, Y. (2005) 'The Indian Madrassahs and the Agenda of Reform'. *Journal of Muslim Minority Affairs*, 25(2): 219–48.

Silverstein, P. A. (2004) *Algeria in France: Transpolitics, Race, and Nation*. Bloomington: Indiana University Press.

Simon, G. M. (2009) 'The soul freed of cares?: Islamic prayer, subjectivity, and the contradictions of moral selfhood in Minangkabau, Indonesia'. *American Ethnologist*, 36(2): 258–75.

Simpson, E. (2008) 'The Changing Perspectives of Three Muslim Men on the Question of Saint-Worship over a 10-Year Period in Gujarat, Western India'. *Modern Asian Studies*, 42(2/3): 377–405.

Simpson, E. (2013) 'Death and the Spirit of Patriarchy in Western India' in Marsden, M. and K. Retsikas (eds) *Articulating Islam: Anthropological Approaches to Muslim Worlds*. New York, London: Springer.

Smith, M. P. (2001) *Transnational Urbanism: Locating Globalization*. Oxford: Blackwell Publishing.

Smith, M. P. and L. E. Guarnizo (eds) (1998) *Transnationalism from Below*. Piscataway, NJ, London: Transaction Publishers.

Soares, B. F. (2005) *Islam and the Prayer Economy: History and Authority in a Malian Town*. Edinburgh: Edinburgh University Press.

Solomos, J. (1993) *Race and Racism in Britain*. Basingstoke: Macmillan.

Spalek, B., and McDonald, L. Z. (2010) 'Terror Crime Prevention: Constructing Muslim Practices and Beliefs as "Anti-Social" and "Extreme" through CONTEST 2'. *Social Policy and Society*, 9(1): 123.

Sperling, David C. (1993) 'Rural *Madrasas* of the Southern Kenya Coast, 1971–92' in L. Brenner, (ed.) *Muslim Identity and Social Change in Sub-Saharan Africa*. London: Hurst & Co.

Stewart, S. B. (2016) *Chinese Muslims and the Global Ummah: Islamic Revival and Ethnic Identity Among the Hui of Qinghai Province*. Oxford: Routledge.

Stocking, G. W. (1991) *Victorian Anthropology*. New York: The Free Press.

Taji-Farouki, S. (1996) *A Fundamental Quest: Hizb al-Tahrir and the search for an Islamic Caliphate*. London: Grey Seal Books.

Tarlo, E. (2010) *Visibly Muslim: Fashion, Politics, Faith*. Oxford, New York: Berg.

Treadwell, J. and J. Garland (2011) 'Masculinity, Marginalization and Violence'. *The British Journal of Criminology*, 51(4): 621–34.

Turner, T. (1993) 'Anthropology and Multiculturalism: What is Anthropology That Multiculturalists Should Be Mindful of It?' *Cultural Anthropology*, 8(4): 411–29.

Van der Veer, P. (1992) 'Playing or Praying: A Sufi Saint's Day in Surat'. *The Journal of Asian Studies*, 51(3): 545–65.

Van Schendel, W. (2009) *A History of Bangladesh*. Cambridge: Cambridge University Press.

Vertigans, S. (2010) 'British Muslims and the UK government's "war on terror" within: evidence of a clash of civilizations or emergent de-civilizing processes?'. *The British Journal of Sociology*, 61(1) (March 2010): 26–44.

Vertovec, S. (1997) 'Three Meanings of "Diaspora," Exemplified among South Asian Religions'. *Diaspora: A Journal of Transnational Studies*, 6(3): 277–99.

Vidino, L. (2010) *The New Muslim Brotherhood in the West*. New York: Colombia University Press.

Visram, R. (2002) *Asians in Britain: 400 Years of History*. London: Pluto Press.

Walter, R. J. (2016) *Coercive Concern: Nationalism, Liberalism, and the Schooling of Muslim Youth*. Stanford, CA: Stanford University Press.

Ward, I. (2006) 'Shabina Begum and the Headscarf Girls'. *Journal of Gender Studies*, 15(2): 119–31.

Watson J. L. (ed.) (1977) *Between Two Cultures: Migrants and Minorities in Britain*. Oxford: Basil Blackwell.

Wemyss, G. (2006) 'The power to tolerate: contests over Britishness and belonging in East London'. *Patterns of Prejudice*, 40(3): 215–36.

Werbner, P. (1979) 'Avoiding the Ghetto: Pakistani migrants and settlement shifts in Manchester'. *New Community*, 7(3): 376–89.

Werbner, P. (1980) 'From rags to riches: Manchester Pakistanis in the textile trade'. *New Community*, 8: 84–95.

Werbner, P. (1987) 'Enclave economies and family firms: Pakistani traders in a British city' in J. Eades (ed.) *Migrants, Workers and the Social Order*. London: Tavistock.

Werbner, P. (1990) *The Migrant Process: Capital, Gifts and Offerings among British Pakistanis*. London, New York: Berg.

Werbner, P. (1991) 'The fiction of unity in ethnic politics: Aspects of representation and the state among British Pakistanis' in P. Werbner and M. Anwar (eds) *Black and Ethnic Leadership in Britain: The cultural dimensions of political action*. London, New York: Routledge.

Werbner, P. (1996) 'Fun Spaces: On Identity and Social Empowerment Among British Pakistanis'. *Theory, Culture & Society*, 13(4): 53–79.

Werbner, P. (2000) 'Divided Loyalties, Empowered Citizenship? Muslims in Britain'. *Citizenship Studies*, 4(3): 307–24.

Werbner, P. (2001) 'The Predicament of Diaspora and Millennial Islam: Reflections on September 11, 2001'. *Ethnicities*, 4: 451–76.

Werbner, P. (2002) *Imagined Diasporas among Manchester Muslims: the public performance of Pakistani transnational identity politics*. Oxford, Santa Fe: School of American Research Press.

Werbner, P. (2004) 'Theorising Complex Diasporas: Purity and Hybridity in the South Asian Public Sphere in Britain'. *Journal of Ethnic and Migration Studies*, 30(5) (September 2004): 895–911.

Werbner, P. and H. Basu (eds) (1998) *Embodying Charisma – Modernity, Locality and the Performance of emotion in Sufi cults*. London: Routledge.

Werbner, P. and T. Modood (eds) (1997) *Debating Cultural Hybridity: Multi-Cultural Identities and the Politics of Anti-Racism*. London: Zed Books.

Wikan, U. (2002) *Generous Betrayal: Politics of Culture in the New Europe*. Chicago, London: Chicago University Press.

Wiktorowicz, Q. (2005) *Radical Islam Rising: Muslim extremism in the West*. Oxford: Rowman and Littlefield.

Wiktorowicz, Q (2006) 'Anatomy of the Salafi Movement'. *Studies in Conflict & Terrorism*, 29(3): 207–39.

Yuval-Davis, N. (2011) *The Politics of Belonging: Intersectional Contestations*. London: Sage Publications.

Zaman, M. Q. (1999) 'Religious Education and the Rhetoric of Reform: The Madrasa in British India and Pakistan'. *Society for Comparative Study of Society and History*, 41(2): 294–323.

Index

Notes are in *italics*.

CPSIA information can be obtained
at www.ICGtesting.com
Printed in the USA
BVHW040304090319
542207BV00006B/55/P